Modern Chemistry

Study Guide
Teacher Edition

HOLT, RINEHART AND WINSTON
A Harcourt Education Company

Orlando • Austin • New York • San Diego • Toronto • London

Copyright © by Holt, Rinehart and Winston

All rights reserved. No part of this publication may be reproduced or transmitted in any form or by any means, electronic or mechanical, including photocopy, recording, or any information storage and retrieval system, without permission in writing from the publisher.

Teachers using MODERN CHEMISTRY may photocopy blackline masters in complete pages in sufficient quantities for classroom use only and not for resale.

HOLT, MODERN CHEMISTRY, and the **"Owl Design"** are trademarks licensed to Holt, Rinehart and Winston, registered in the United States of America and/or other jurisdictions.

Printed in the United States of America

If you have received these materials as examination copies free of charge, Holt, Rinehart and Winston retains title to the materials and they may not be resold. Resale of examination copies is strictly prohibited.

Possession of this publication in print format does not entitle users to convert this publication, or any portion of it, into electronic format.

ISBN 0-03-036778-6

6 7 8 9   170   07

Contents

1 Matter and Change
- Section 1 Chemistry Is a Physical Science ... 1
- Section 2 Matter and Its Properties ... 3
- Section 3 Elements ... 5
- Chapter 1 Mixed Review ... 7

2 Measurements and Calculations
- Section 1 Scientific Method ... 9
- Section 2 Units of Measurement ... 11
- Section 3 Using Scientific Measurements ... 12
- Chapter 2 Mixed Review ... 15

3 Atoms: The Building Blocks of Matter
- Section 1 The Atom: From Philosophical Idea to Scientific Theory ... 17
- Section 2 The Structure of the Atom ... 19
- Section 3 Counting Atoms ... 21
- Chapter 3 Mixed Review ... 23

4 Arrangement of Electrons in Atoms
- Section 1 The Development of a New Atomic Model ... 25
- Section 2 The Quantum Model of the Atom ... 27
- Section 3 Electron Configurations ... 29
- Chapter 4 Mixed Review ... 31

5 The Periodic Law
- Section 1 History of the Periodic Table ... 33
- Section 2 Electron Configuration and the Periodic Table ... 35
- Section 3 Electron Configuration and Periodic Properties ... 37
- Chapter 5 Mixed Review ... 39

6 Chemical Bonding
- Section 1 Introduction to Chemical Bonding ... 41
- Section 2 Covalent Bonding and Molecular Compounds ... 43
- Section 3 Ionic Bonding and Ionic Compounds ... 45
- Section 4 Metallic Bonding ... 47
- Section 5 Molecular Geometry ... 49
- Chapter 6 Mixed Review ... 51

7 Chemical Formulas and Chemical Compounds
- Section 1 Chemical Names and Formulas ... 53
- Section 2 Oxidation Numbers ... 55
- Section 3 Using Chemical Formulas ... 57
- Section 4 Determining Chemical Formulas ... 59
- Chapter 7 Mixed Review ... 61

8 Chemical Equations and Reactions
Section 1 Describing Chemical Reactions 65
Section 2 Types of Chemical Reactions 67
Section 3 Activity Series of the Elements 69
Chapter 8 Mixed Review 71

9 Stoichiometry
Section 1 Introduction to Stoichiometry 73
Section 2 Ideal Stoichiometric Calculations 75
Section 3 Limiting Reactants and Percentage Yield 77
Chapter 9 Mixed Review 79

10 States of Matter
Section 1 Kinetic-Molecular Theory of Matter 81
Section 2 Liquids 83
Section 3 Solids 85
Section 4 Changes of State 87
Section 5 Water 89
Chapter 10 Mixed Review 91

11 Gases
Section 1 Gases and Pressure 93
Section 2 The Gas Laws 95
Section 3 Gas Volumes and the Ideal Gas Law 97
Section 4 Diffusion and Effusion 99
Chapter 11 Mixed Review 100

12 Solutions
Section 1 Types of Mixtures 103
Section 2 The Solution Process 105
Section 3 Concentration of Solutions 107
Chapter 12 Mixed Review 109

13 Ions in Aqueous Solutions and Colligative Properties
Section 1 Compounds in Aqueous Solutions 111
Section 2 Colligative Properties of Solutions 113
Chapter 13 Mixed Review 115

14 Acids and Bases
Section 1 Properties of Acids and Bases 117
Section 2 Acid-Base Theories 119
Section 3 Acid-Base Reactions 121
Chapter 14 Mixed Review 123

15 Acid-Base Titration and pH
Section 1 Aqueous Solutions and the Concept of pH 125
Section 2 Determining pH and Titrations 127
Chapter 15 Mixed Review 129

16 Reaction Energy
- **Section 1** Thermochemistry 131
- **Section 2** Driving Force of Reactions 133
- **Chapter 16** Mixed Review 135

17 Reaction Kinetics
- **Section 1** The Reaction Process....................... 137
- **Section 2** Reaction Rates 139
- **Chapter 17** Mixed Review 141

18 Chemical Equilibrium
- **Section 1** The Nature of Chemical Equilibrium......... 143
- **Section 2** Shifting Equilibrium 145
- **Section 3** Equilibria of Acids, Bases, and Salts...... 147
- **Section 4** Solubility Equilibrium..................... 149
- **Chapter 18** Mixed Review 151

19 Oxidation-Reduction Reactions
- **Section 1** Oxidation and Reduction 153
- **Section 2** Balancing Redox Equations.................. 155
- **Section 3** Oxidizing and Reducing Agents.............. 157
- **Chapter 19** Mixed Review 159

20 Electrochemistry
- **Section 1** Introduction to Electrochemistry 161
- **Section 2** Voltaic Cells 163
- **Section 3** Electrolytic Cells 165
- **Chapter 20** Mixed Review 167

21 Nuclear Chemistry
- **Section 1** The Nucleus 169
- **Section 2** Radioactive Decay 171
- **Section 3** Nuclear Radiation.......................... 173
- **Section 4** Nuclear Fission and Nuclear Fusion 175
- **Chapter 21** Mixed Review 177

22 Organic Chemistry
- **Section 1** Organic Compounds.......................... 179
- **Section 2** Hydrocarbons 181
- **Section 3** Functional Groups 183
- **Section 4** Organic Reactions 185
- **Chapter 22** Mixed Review 187

23 Biological Chemistry
- **Section 1** Carbohydrates and Lipids 189
- **Section 2** Amino Acids and Proteins 191
- **Section 3** Metabolism 193
- **Section 4** Nucleic Acids 195
- **Chapter 23** Mixed Review 197

Name _____ Date _____ Class _____

CHAPTER 1 REVIEW
Matter and Change

SECTION 1

SHORT ANSWER Answer the following questions in the space provided.

1. __a__ Technological development of a chemical product often
   - (a) lags behind basic research on the same substance.
   - (b) does not involve chance discoveries.
   - (c) is driven by curiosity.
   - (d) is done for the sake of learning something new.

2. __d__ The primary motivation behind basic research is to
   - (a) develop new products.
   - (b) make money.
   - (c) understand an environmental problem.
   - (d) gain knowledge.

3. __a__ Applied research is designed to
   - (a) solve a particular problem.
   - (b) satisfy curiosity.
   - (c) gain knowledge.
   - (d) learn for the sake of learning.

4. __b__ Chemistry is usually classified as
   - (a) a biological science.
   - (b) a physical science.
   - (c) a social science.
   - (d) a computer science.

5. Define the six major branches of chemistry.

   organic chemistry—the study of carbon-containing compounds

   inorganic chemistry—the study of non-organic substances

   physical chemistry—the study of properties of matter, changes that occur in matter,

   and the relationships between matter and energy

   analytical chemistry—the identification of the composition of materials

   biochemistry—the study of the chemistry of living things

   theoretical chemistry—the use of mathematics and computers to design and

   predict the properties of new compounds

MODERN CHEMISTRY   MATTER AND CHANGE   1
Copyright © by Holt, Rinehart and Winston. All rights reserved.

SECTION 1 continued

6. For each of the following types of chemical investigations, determine whether the investigation is *basic research*, *applied research*, or *technological development*. More than one choice may apply.

    __basic research__ a. A laboratory in a major university surveys all the reactions involving bromine.

    __applied research/ technical development__ b. A pharmaceutical company explores a disease in order to produce a better medicine.

    __applied research__ c. A scientist investigates the cause of the ozone hole to find a way to stop the loss of the ozone layer.

    __applied research/ technical development__ d. A pharmaceutical company discovers a more efficient method of producing a drug.

    __applied research/ technical development__ e. A chemical company develops a new biodegradable plastic.

    __applied research__ f. A laboratory explores the use of ozone to inactivate bacteria in a drinking-water system.

7. Give examples of two different instruments routinely used in chemistry.
    Answers may include any type of balance and any type of microscope.

8. What are microstructures?
    things too small to be seen with the unaided eye

9. What is a chemical?
    a substance with a definite composition

10. What is chemistry?
    the study of the composition, properties, and interactions of matter

Name _____ Date _____ Class _____

CHAPTER 1 REVIEW
Matter and Change

SECTION 2

SHORT ANSWER Answer the following questions in the space provided.

1. Classify each of the following as a *homogeneous* or *heterogeneous* substance.

    - heterogeneous — **a.** iron ore
    - homogeneous — **b.** quartz
    - heterogeneous — **c.** granite
    - homogeneous — **d.** energy drink
    - heterogeneous — **e.** oil-and-vinegar salad dressing
    - homogeneous — **f.** salt
    - homogeneous — **g.** rainwater
    - homogeneous — **h.** nitrogen

2. Classify each of the following as a *physical* or *chemical* change.

    - physical — **a.** ice melting
    - chemical — **b.** paper burning
    - chemical — **c.** metal rusting
    - physical — **d.** gas pressure increasing
    - physical — **e.** liquid evaporating
    - chemical — **f.** food digesting

3. Compare a physical change with a chemical change.

    A chemical change involves a rearrangement of the atoms of different elements in a substance and the formation substances with different physical properties. A physical change can occur in properties such as the state or shape of a substance, but it will not affect the composition of that substance.

MODERN CHEMISTRY   MATTER AND CHANGE   3

SECTION 2 continued

4. Compare and contrast each of the following terms:

   a. *mass* and *matter*

   Mass is a measure of the amount of matter. Matter is anything that has mass and takes up space.

   b. *atom* and *compound*

   All matter is composed of atoms, which are the smallest units of an element that retain the properties of that element. Atoms can come together to form compounds.

   c. *physical property* and *chemical property*

   Physical properties are characteristics such as color, density, melting point, and boiling point that can be measured without changing the identity of the substance. Chemical properties relate to how a substance interacts with another substance to form a different substance.

   d. *homogeneous mixture* and *heterogeneous mixture*

   A homogeneous mixture has a uniform composition. A heterogeneous mixture is not uniform.

5. Using circles to represent particles, draw a diagram that compares the arrangement of particles in the solid, liquid, and gas states.

   Solid   Liquid   Gas

6. How is energy involved in chemical and physical changes?

   Energy is either absorbed or given off in all chemical and physical changes, but it is neither created nor destroyed. It simply assumes a different form, or it is moved from one place to another.

CHAPTER 1 REVIEW
Matter and Change

SECTION 3

SHORT ANSWER Answer the following questions in the space provided.

1. A horizontal row of elements in the periodic table is called a(n) ____period____.

2. The symbol for the element in Period 2, Group 13, is ____B____.

3. Elements that are good conductors of heat and electricity are ____metals____.

4. Elements that are poor conductors of heat and electricity are ____nonmetals____.

5. A vertical column of elements in the periodic table is called a(n) ____group, or family____.

6. The ability of a substance to be hammered or rolled into thin sheets is called ____malleability____.

7. Is an element that is soft and easy to cut cleanly with a knife likely to be a metal or a nonmetal? ____metal____

8. The elements in Group 18, which are generally unreactive, are called ____noble gases____.

9. At room temperature, most metals are ____solids____.

10. Name three characteristics of most nonmetals.
    They are brittle, are poor conductors of heat and electricity, and have low boiling points.

11. Name three characteristics of metals.
    They are malleable, ductile, and good conductors of heat and electricity, and they have a metallic (shiny) luster.

12. Name three characteristics of most metalloids.
    They are semiconductors of electricity, solid at room temperature, and less malleable than metals.

13. Name two characteristics of noble gases.
    They are in the gas state at room temperature and are generally unreactive.

SECTION 3 continued

14. What do elements of the same group in the periodic table have in common?

Elements of the same group share similar chemical properties.

15. Within the same period of the periodic table, how do the properties of elements close to each other compare with the properties of elements far from each other?

The properties of elements that are close to each other in the same period tend to be more similar than the properties of elements that are far apart. Physical and chemical properties change somewhat regularly across a period.

16. You are trying to manufacture a new material, but you would like to replace one of the elements in your new substance with another element that has similar chemical properties. How would you use the periodic table to choose a likely substitute?

You would consider an element of the same vertical column, or group, because elements in the same group have similar chemical properties.

17. What is the difference between a family of elements and elements in the same period?

***Family* is another name for *group*, or elements in the same vertical column.**

Elements in the same period are in the same horizontal row.

18. Complete the table below by filling in the spaces with correct names or symbols.

Name of element	Symbol of element
Aluminum	Al
Calcium	Ca
Manganese	Mn
Nickel	Ni
Potassium	K
Cobalt	Co
Silver	Ag
Hydrogen	H

Name _____ Date _____ Class _____

CHAPTER 1 REVIEW
Matter and Change

MIXED REVIEW

SHORT ANSWER Answer the following questions in the space provided.

1. Classify each of the following as a *homogeneous* or *heterogeneous* substance.

   __homogeneous__ a. sugar   __homogeneous__ d. plastic wrap

   __homogeneous__ b. iron filings   __heterogeneous__ e. cement sidewalk

   __heterogeneous__ c. granola bar

2. For each type of investigation, select the most appropriate branch of chemistry from the following choices: *organic chemistry, analytical chemistry, biochemistry, theoretical chemistry*. More than one branch may be appropriate.

   __analytical chemistry__ a. A forensic scientist uses chemistry to find information at the scene of a crime.

   __theoretical chemistry/ biochemistry__ b. A scientist uses a computer model to see how an enzyme will function.

   __biochemistry__ c. A professor explores the reactions that take place in a human liver.

   __organic chemistry__ d. An oil company scientist tries to design a better gasoline.

   __analytical chemistry__ e. An anthropologist tries to find out the nature of a substance in a mummy's wrap.

   __biochemistry/ analytical chemistry__ f. A pharmaceutical company examines the protein on the coating+ of a virus.

3. For each of the following types of chemical investigations, determine whether the investigation is *basic research, applied research,* or *technological development*. More than one choice may apply.

   __basic research__ a. A university plans to map all the genes on human chromosomes.

   __applied research__ b. A research team intends to find out why a lake remains polluted to try to find a way to clean it up.

   __applied research/ technological development__ c. A science teacher looks for a solvent that will allow graffiti to be removed easily.

   __basic research/ applied research__ d. A cancer research institute explores the chemistry of the cell.

   __basic research__ e. A professor explores the toxic compounds in marine animals.

Name _____ Date _____ Class _____

MIXED REVIEW continued

4. Use the periodic table to identify the name, group number, and period number of the following elements:

   __chlorine, Group 17, Period 3__ a. Cl
   __magnesium, Group 2, Period 3__ b. Mg
   __tungsten, Group 6, Period 6__ c. W
   __iron, Group 8, Period 4__ d. Fe
   __tin, Group 14, Period 5__ e. Sn

5. What is the difference between extensive and intensive properties?
   Extensive properties depend on the amount of matter present; intensive properties do not.

6. Consider the burning of gasoline and the evaporation of gasoline. Which process represents a chemical change and which represents a physical change? Explain your answer.
   The burning of gasoline represents a chemical change because the gasoline is being changed into substances with different identities. Evaporation involves a physical change; the identity of gasoline remains unchanged.

7. Describe the difference between a heterogeneous mixture and a homogeneous mixture, and give an example of each.
   A heterogeneous mixture, such as blood, is made of components with different physical properties. A homogeneous mixture, such as stainless steel, has a single set of physical properties.

8. Construct a concept map that includes the following terms: *atom, element, compound, pure substance, mixture, homogeneous,* and *heterogeneous*.

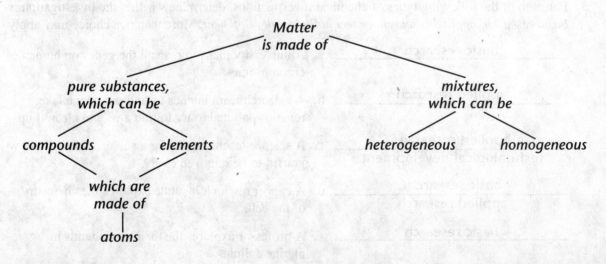

8 MATTER AND CHANGE

Name _____ Date _____ Class _____

CHAPTER 2 REVIEW
Measurements and Calculations

SECTION 1

SHORT ANSWER Answer the following questions in the space provided.

1. Determine whether each of the following is an example of *observation and data*, a *theory*, a *hypothesis*, a *control*, or a *model*.

    __observation and data__ a. A research team records the rainfall in inches per day in a prescribed area of the rain forest. The square footage of vegetation and relative plant density per square foot are also measured.

    __observation and data__ b. The intensity, duration, and time of day of the precipitation are noted for each precipitation episode. The types of vegetation in the area are recorded and classified.

    __control__ c. The information gathered is compared with the data on the average precipitation and the plant population collected over the last 10 years.

    __hypothesis__ d. The information gathered by the research team indicates that rainfall has decreased significantly. They propose that deforestation is the primary cause of this phenomenon.

2. "When 10.0 g of a white, crystalline sugar are dissolved in 100. mL of water, the solution is observed to freeze at $-0.54°C$, not $0.0°C$. The system is denser than pure water." Which parts of these statements represent quantitative information, and which parts represent qualitative information?

    Quantitative values include the mass of sugar, volume of water, and observed

    freezing point. Qualitative properties are the color and state of the sugar and the

    claim of greater density.

3. Compare and contrast a model with a theory.

    Theories are broad generalizations used to explain observations. Models are

    a physical object used to illustrate or explain complex concepts or an explanation of

    how phenomena occur and how data and events are related.

MODERN CHEMISTRY MEASUREMENTS AND CALCULATIONS

SECTION 1 continued

4. Evaluate the models shown below. Describe how the models resemble the objects they represent and how they differ from the objects they represent.

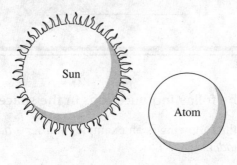

The model of the sun accurately shows that the sun is round and has a fiery surface, but the model is much smaller than the real sun and does not show the sun's composition. The model of an atom accurately shows that an atom is a particle, but the model is much larger than a real atom and does not depict an atom's composition or shape.

5. __c__ How many different variables are represented in the two graphs shown below?
   a. one   b. two   c. three   d. four

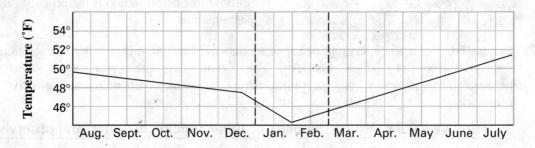

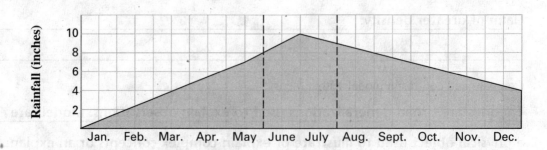

Name _____ Date _____ Class _____

CHAPTER 2 REVIEW
Measurements and Calculations

SECTION 2

SHORT ANSWER Answer the following questions in the space provided.

1. Complete the following conversions:

   a. 100 mL = _____0.1_____ L

   b. 0.25 g = _____25_____ cg

   c. 400 cm^3 = _____0.4_____ L

   d. 400 cm^3 = _____0.0004_____ m^3

2. For each measuring device shown below, identify the quantity measured and tell when it would remain constant and when it would vary.

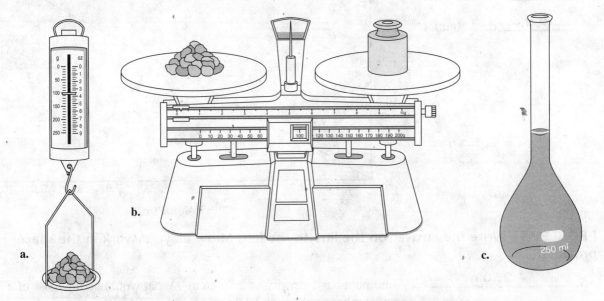

Device **a** measures weight (the effect of the gravitational force on mass), which changes with location on Earth and when measured on a different planet or moon. Device **b** measures mass, which does not change with location because gravity affects both the measured body and the mass standard equally. Device **c** measures volume of a liquid, which changes slightly with temperature and pressure. Weight, mass, and volume do not change with the shape of the object.

MODERN CHEMISTRY MEASUREMENTS AND CALCULATIONS **11**

Name _____ Date _____ Class _____

SECTION 2 continued

3. Use the data found in **Table 4** on page 38 of the text to answer the following questions:

   _____sink_____ a. If ice were denser than liquid water at 0°C, would it float or sink in water?

   _____kerosene_____ b. Water and kerosene do not dissolve readily in one another. If the two are mixed, they quickly separate into layers. Which liquid floats on top?

   _____mercury_____ c. The other liquids in **Table 4** that do not dissolve in water are gasoline, turpentine, and mercury. Which of these liquids would settle to the bottom when mixed with water?

4. Use the graph of the density of aluminum below to determine the approximate mass of aluminum samples with the following volumes.

   __22 g__ a. 8.0 mL
   __4 g__ b. 1.50 mL
   __20 g__ c. 7.25 mL
   __9 g__ d. 3.50 mL

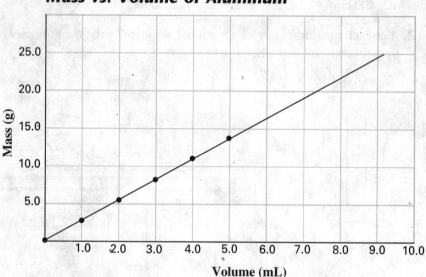

PROBLEMS Write the answer on the line to the left. Show all your work in the space provided.

5. _____27.0 g_____ Aluminum has a density of 2.70 g/cm^3. What would be the mass of a sample whose volume is 10.0 cm^3?

6. _____14 cm_____ A certain piece of copper wire is determined to have a mass of 2.00 g per meter. How many centimeters of the wire would be needed to provide 0.28 g of copper?

Name ______________________ Date ______________ Class ______________

CHAPTER 2 REVIEW

Measurements and Calculations

SECTION 3

SHORT ANSWER **Answer the following questions in the space provided.**

1. Report the number of significant figures in each of the following values:

3 **a.** 0.002 37 g

4 **b.** 0.002 037 g

3 **c.** 350. J

2 **d.** 64 mL

2 **e.** 1.3×10^2 cm

3 **f.** 1.30×10^2 cm

2. Write the value of the following operations using scientific notation.

10^{-1} **a.** $\frac{10^3 \times 10^{-6}}{10^{-2}}$

4×10^{-2} **b.** $\frac{8 \times 10^3}{2 \times 10^5}$

4.3×10^4 **c.** $3 \times 10^3 + 4.0 \times 10^4$

3. The following data are given for two variables, A and B:

A	*B*
18	2
9	4
6	6
3	12

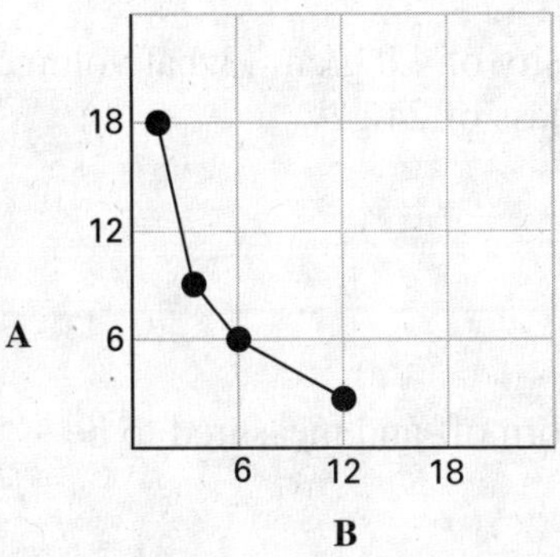

a. In the graph provided, plot the data.

inversely proportional **b.** Are *A* and *B* directly or inversely proportional?

No **c.** Do the data points form a straight line?

$A \times B = k$ **d.** Which equation fits the relationship shown by the data?

$\frac{A}{B} = k$ (a constant) or $A \times B = k$ (a constant)

36 **e.** What is the value of *k*?

SECTION 3 continued

4. Carry out the following calculations. Express each answer to the correct number of significant figures and use the proper units.

   __40.0 m__ a. 37.26 m + 2.7 m + 0.0015 m =

   __2000 mL or 2 L__ b. 256.3 mL + 2 L + 137 mL =

   __151 mL__ c. $\dfrac{300.\ \text{kPa} \times 274.57\ \text{mL}}{547\ \text{kPa}} =$

   __100 mL__ d. $\dfrac{346\ \text{mL} \times 200\ \text{K}}{546.4\ \text{K}} =$

5. Round the following measurements to three significant figures.

   __22.8 g__ a. 22.77 g

   __14.6 m__ b. 14.62 m

   __9.31 L__ c. 9.3052 L

   __87.6 cm__ d. 87.55 cm

   __30.2 g__ e. 30.25 g

PROBLEMS Write the answer on the line to the left. Show all your work in the space provided.

6. A pure solid at a fixed temperature has a constant density. We know that
   $$\text{density} = \dfrac{\text{mass}}{\text{volume}} \text{ or } D = \dfrac{m}{V}.$$

   __directly proportional__ a. Are mass and volume directly proportional or inversely proportional for a fixed density?

   __6.0 cm³__ b. If a solid has a density of 4.0 g/cm³, what volume of the solid has a mass of 24 g?

7. A crime-scene tape has a width of 13.8 cm. A long strip of it is torn off and measured to be 56 m long.

   __5600 cm__ a. Convert 56 m into centimeters.

   __7.7 × 10⁴ cm²__ b. What is the area of this rectangular strip of tape, in cm²?

CHAPTER 2 REVIEW
Measurements and Calculations

MIXED REVIEW

SHORT ANSWER Answer the following questions in the space provided.

1. Match the description on the right to the most appropriate quantity on the left.

    __d__ 2 m^3 (a) mass of a small paper clip

    __a__ 0.5 g   (b) length of a small paper clip

    __f__ 0.5 kg   (c) length of a stretch limousine

    __e__ 600 cm^2 (d) volume of a refrigerator compartment

    __b__ 20 mm   (e) surface area of the cover of this workbook

                           (f) mass of a jar of peanut butter

2. __a__ A measured quantity is said to have good accuracy if

    (a) it agrees closely with the accepted value.
    (b) repeated measurements agree closely.
    (c) it has a small number of significant figures.
    (d) all digits in the value are significant.

3. A certain sample with a mass of 4.00 g is found to have a volume of 7.0 mL. To calculate the density of the sample, a student entered 4.00 ÷ 7.0 on a calculator. The calculator display shows the answer as 0.571429.

    __Yes__ a. Is the setup for calculating density correct?

    __2__ b. How many significant figures should the answer contain?

4. It was shown in the text that in a value such as 4000 g, the precision of the number is uncertain. The zeros may or may not be significant.

    __1__ a. Suppose that the mass was determined to be 4000 g. How many significant figures are present in this measurement?

    __4.00×10^3 g__ b. Suppose you are told that the mass lies somewhere between 3950 and 4050 g. Use scientific notation to report the value, showing an appropriate number of significant figures.

5. If you divide a sample's mass by its density, what are the resulting units?

    Volume units: for example, $\dfrac{g}{(g/mL)} = mL$

Name _____ Date _____ Class _____

MIXED REVIEW continued

6. Three students were asked to determine the volume of a liquid by a method of their choosing. Each performed three trials. The table below shows the results. The actual volume of the liquid is 24.8 mL.

	Trial 1 (mL)	Trial 2 (mL)	Trial 3 (mL)
Student A	24.8	24.8	24.4
Student B	24.2	24.3	24.3
Student C	24.6	24.8	25.0

__Student C__ a. Considering the average of all three trials, which student's measurements show the greatest accuracy?

__Student B__ b. Which student's measurements show the greatest precision?

PROBLEMS Write the answer on the line to the left. Show all your work in the space provided.

7. __2.0×10^2 g__ A single atom of platinum has a mass of 3.25×10^{-22} g. What is the mass of 6.0×10^{23} platinum atoms?

8. A sample thought to be pure lead occupies a volume of 15.0 mL and has a mass of 160.0 g.

__10.7 g/mL__ a. Determine its density.

__No__ b. Is the sample pure lead? (Refer to **Table 4** on page 38 of the text.)

__6.0%__ c. Determine the percentage error, based on the accepted value for the density of lead.

CHAPTER 3 REVIEW
Atoms: The Building Blocks of Matter

SECTION 1

SHORT ANSWER Answer the following questions in the space provided.

1. Why is Democritus's view of matter considered only an idea, while Dalton's view is considered a theory?

   Democritus's idea of matter does not relate atoms to a measurable property, while Dalton's theory can be tested through quantitative experimentation.

2. Give an example of a chemical or physical process that illustrates the law of conservation of mass.

   A glass of ice cubes will have the same mass when the ice has completely melted into liquid water, even though its volume will change. (Accept any reasonable process.)

3. State two principles from Dalton's atomic theory that have been revised as new information has become available.

   Atoms are divisible into smaller particles called subatomic particles. A given element can have atoms with different masses, called isotopes.

4. The formation of water according to the equation

   $$2H_2 + O_2 \rightarrow 2H_2O$$

   shows that 2 molecules (made of 4 atoms) of hydrogen and 1 molecule (made of 2 atoms) of oxygen produce 2 molecules of water. The total mass of the product, water, is equal to the sum of the masses of each of the reactants, hydrogen and oxygen. What parts of Dalton's atomic theory are illustrated by this reaction? What law does this reaction illustrate?

   Atoms cannot be subdivided, created, or destroyed. Also, atoms of different elements combine in simple, whole-number ratios to form compounds. The reaction also illustrates the law of conservation of mass.

Name _____ Date _____ Class _____

SECTION 1 continued

PROBLEMS Write the answer on the line to the left. Show all your work in the space provided.

5. _____16 g_____ If 3 g of element C combine with 8 g of element D to form compound CD, how many grams of D are needed to form compound CD_2?

6. A sample of baking soda, $NaHCO_3$, *always* contains 27.37% by mass of sodium, 1.20% of hydrogen, 14.30% of carbon, and 57.14% of oxygen.

   a. Which law do these data illustrate?
   the law of definite proportions

   b. State the law.
   A chemical compound contains the same elements in exactly the same proportions by mass regardless of the sample or the source of the compound.

7. Nitrogen and oxygen combine to form several compounds, as shown by the following table.

Compound	Mass of nitrogen that combines with 1 g oxygen (g)
NO	1.70
NO_2	0.85
NO_4	0.44

   Calculate the ratio of the masses of nitrogen in each of the following:

   __2.0__ a. $\dfrac{NO}{NO_2}$ __2.0__ b. $\dfrac{NO_2}{NO_4}$ __4.0__ c. $\dfrac{NO}{NO_4}$

   d. Which law do these data illustrate?
   the law of multiple proportions

CHAPTER 3 REVIEW
Atoms: The Building Blocks of Matter

SECTION 2

SHORT ANSWER Answer the following questions in the space provided.

1. In cathode-ray tubes, the cathode ray is emitted from the negative electrode, which is called the __cathode__.

2. The smallest unit of an element that can exist either alone or in molecules containing the same or different elements is the __atom__.

3. A positively charged particle found in the nucleus is called a(n) __proton__.

4. A nuclear particle that has no electrical charge is called a(n) __neutron__.

5. The subatomic particles that are least massive and most massive, respectively, are the __electron__ and __neutron__.

6. A cathode ray produced in a gas-filled tube is deflected by a magnetic field. A wire carrying an electric current can be pulled by a magnetic field. A cathode ray is deflected away from a negatively charged object. What property of the cathode ray is shown by these phenomena?
   The particles that compose cathode rays are negatively charged.

7. How would the electrons produced in a cathode-ray tube filled with neon gas compare with the electrons produced in a cathode-ray tube filled with chlorine gas?
   The electrons produced from neon gas and chlorine gas would behave in the same way because electrons do not differ from element to element.

8. a. Is an atom positively charged, negatively charged, or neutral?
      Atoms are neutral.
   b. Explain how an atom can exist in this state.
      Atoms consist of a positively charged nucleus, made up of protons and neutrons, that is surrounded by a negatively charged electron cloud. The positive and negative charges combine to form a net neutral charge.

Name _____ Date _____ Class _____

SECTION 2 continued

9. Below are illustrations of two scientists' conceptions of the atom. Label the electrons in both illustrations with a − sign and the nucleus in the illustration to the right with a + sign. On the lines below the figures, identify which illustration was believed to be correct before Rutherford's gold-foil experiment and which was believed to be correct after Rutherford's gold-foil experiment.

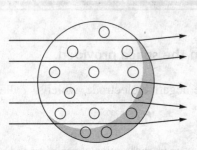
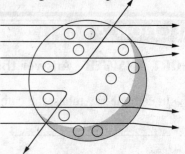

(Students should place a − sign inside all circles.)

a. __before Rutherford's experiment__

(Students should place a + sign in the center and a − sign inside all circles.)

b. __after Rutherford's experiment__

10. In the space provided, describe the locations of the subatomic particles in the labeled model of an atom of nitrogen below, and give the charge and relative mass of each particle.

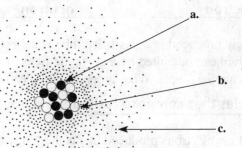

a. proton

The proton, a positive and relatively massive particle, should be located in the nucleus.

b. neutron

The neutron, a neutral and relatively massive particle, should be located in the nucleus.

c. electron (a possible location of this particle)

The electron, a negative particle with a low mass, should be located in the cloud surrounding the nucleus.

CHAPTER 3 REVIEW
Atoms: The Building Blocks of Matter

SECTION 3

SHORT ANSWER Answer the following questions in the space provided.

1. Explain the difference between the *mass number* and the *atomic number* of a nuclide.

   Mass number is the total number of protons and neutrons in the nucleus of an

   isotope. Atomic number is the total number of protons in the nucleus of

   each atom of an element.

2. Why is it necessary to use the average atomic mass of all isotopes, rather than the mass of the most commonly occurring isotope, when referring to the atomic mass of an element?

   Elements rarely occur as only one isotope; rather, they exist as mixtures

   of different isotopes of various masses. Using a weighted average atomic mass, you

   can account for the less common isotopes.

3. How many particles are in 1 mol of carbon? 1 mol of lithium? 1 mol of eggs? Will 1 mol of each of these substances have the same mass?

   There are 6.022×10^{23} particles in 1 mol of each of these substances. One mole of

   one substance will not necessarily have the same mass as one mole of another

   substance.

4. Explain what happens to each of the following as the atomic masses of the elements in the periodic table increase:

   a. the number of protons

   increases

   b. the number of electrons

   increases

   c. the number of atoms in 1 mol of each element

   stays the same

SECTION 3 continued

5. Use a periodic table to complete the following chart:

Element	Symbol	Atomic number	Mass number
Europium-151	$^{151}_{63}$Eu	63	151
Silver-109	$^{109}_{47}$Ag	47	109
Tellurium-128	$^{128}_{52}$Te	52	128

6. List the number of protons, neutrons, and electrons found in zinc-66.

   __30__ protons

   __36__ neutrons

   __30__ electrons

PROBLEMS Write the answer on the line to the left. Show all your work in the space provided.

7. _____32.00 g_____ What is the mass in grams of 2.000 mol of oxygen atoms?

8. _____3.706 mol_____ How many moles of aluminum exist in 100.0 g of aluminum?

9. __1.994 × 10^{24} atoms__ How many atoms are in 80.45 g of magnesium?

10. __1.993 × 10^{-21} g__ What is the mass in grams of 100 atoms of the carbon-12 isotope?

CHAPTER 3 REVIEW
Atoms: The Building Blocks of Matter

MIXED REVIEW

SHORT ANSWER Answer the following questions in the space provided.

1. The element boron, B, has an atomic mass of 10.81 amu according to the periodic table. However, no single atom of boron has a mass of exactly 10.81 amu. How can you explain this difference?

   The periodic table reports the average atomic mass, which is a weighted average of

   all isotopes of boron.

2. How did the outcome of Rutherford's gold-foil experiment indicate the existence of a nucleus?

   A few alpha particles rebounded and therefore must have "hit" a dense bundle of

   matter. Because such a small percentage of particles were redirected, he reasoned

   that this clump of matter, called the nucleus, must occupy only a small fraction of

   the atom's total space.

3. Ibuprofen, $C_{13}H_{18}O_2$, that is manufactured in Michigan contains 75.69% by mass carbon, 8.80% hydrogen, and 15.51% oxygen. If you buy some ibuprofen for a headache while you are on vacation in Germany, how do you know that it has the same percentage composition as the ibuprofen you buy at home?

   The law of definite proportions states that a chemical compound contains the same

   elements in exactly the same proportions by mass regardless of the site of the

   sample or the source of the compound.

4. Complete the following chart, using the atomic mass values from the periodic table:

Compound	Mass of Fe (g)	Mass of O (g)	Ratio of O:Fe
FeO	55.85	16.00	0.2865
Fe_2O_3	111.70	48.00	0.4297
Fe_3O_4	167.55	64.00	0.3820

MIXED REVIEW continued

5. Complete the following table:

Element	Symbol	Atomic number	Mass number	Number of protons	Number of neutrons	Number of electrons
Sodium	Na	11	22	11	11	11
Fluorine	F	9	19	9	10	9
Bromine	Br	35	80	35	45	35
Calcium	Ca	20	40	20	20	20
Hydrogen	H	1	1	1	0	1
Radon	Rn	86	222	86	136	86

PROBLEMS Write the answer on the line to the left. Show all your work in the space provided.

6. __1.51×10^{24} atoms__ a. How many atoms are there in 2.50 mol of hydrogen?

   __1.51×10^{24} atoms__ b. How many atoms are there in 2.50 mol of uranium?

   __4.65 mol__ c. How many moles are present in 107 g of sodium?

7. A certain element exists as three natural isotopes, as shown in the table below.

Isotope	Mass (amu)	Percent natural abundance	Mass number
1	19.99244	90.51	20
2	20.99395	0.27	21
3	21.99138	9.22	22

   __20 amu__ Calculate the average atomic mass of this element.

CHAPTER 4 REVIEW
Arrangement of Electrons in Atoms

SECTION 1

SHORT ANSWER Answer the following questions in the space provided.

1. In what way does the photoelectric effect support the particle theory of light?

   In order for an electron to be ejected from a metal surface, the electron must be struck by a single photon with at least the minimum energy needed to knock the electron loose.

2. What is the difference between the ground state and the excited state of an atom?

   The ground state is the lowest energy state of the atom. When the atom absorbs energy, it can move to a higher energy state, or excited state.

3. Under what circumstances can an atom emit a photon?

   A photon is emitted when an atom moves from an excited state to its ground state or to a lower-energy excited state.

4. How can the energy levels of the atom be determined by measuring the light emitted from an atom?

   When an atom loses energy, it falls from a higher energy state to a lower energy state. The frequency of the emitted light, observed in an element's line-emission spectrum, may be measured. The energy of each transition is calculated using the equation $E = h\nu$, where ν is the frequency of each of the lines in the element's line-emission spectrum. From the analysis of these results, the energy levels of an atom atom of each element may be determined.

5. Why does electromagnetic radiation in the ultraviolet region represent a larger energy transition than does radiation in the infrared region?

   Energy is proportional to frequency, and ultraviolet radiation has a higher frequency than infrared radiation. To produce ultraviolet radiation, electrons must drop to lower energy levels than they do to produce infrared radiation.

SECTION 1 continued

6. Which of the waves shown below has the higher frequency? (The scale is the same for each drawing.) Explain your answer.

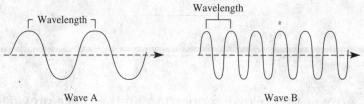

Wave A   Wave B

Wave B has the higher frequency. Wavelength is inversely proportional to

frequency, so as the wavelength decreases, its frequency increases.

7. How many different photons of radiation were emitted from excited helium atoms to form the spectrum shown below? Explain your answer.

Spectrum for helium

Six different photons were emitted. Each time an excited helium atom falls back

from an excited state to its ground state or to a lower energy state, it emits a

photon of radiation that shows up as this specific line-emission spectrum. There are

six lines in this helium spectrum.

PROBLEMS Write the answer on the line to the left. Show all your work in the space provided.

8. _____9.7×10^{14} Hz_____ What is the frequency of light that has a wavelength of 310 nm?

9. _____9.4×10^{9} m_____ What is the wavelength of electromagnetic radiation if its frequency is 3.2×10^{-2} Hz?

CHAPTER 4 REVIEW
Arrangement of Electrons in Atoms

SECTION 2

SHORT ANSWER Answer the following questions in the space provided.

1. __d__ How many quantum numbers are used to describe the properties of electrons in atomic orbitals?
   - (a) 1
   - (b) 2
   - (c) 3
   - (d) 4

2. __a__ A spherical electron cloud surrounding an atomic nucleus would best represent
   - (a) an *s* orbital.
   - (b) a *p* orbital.
   - (c) a combination of two different *p* orbitals.
   - (d) a combination of an *s* and a *p* orbital.

3. __a__ How many electrons can an energy level of $n = 4$ hold?
   - (a) 32
   - (b) 24
   - (c) 8
   - (d) 6

4. __c__ How many electrons can an energy level of $n = 2$ hold?
   - (a) 32
   - (b) 24
   - (c) 8
   - (d) 6

5. __c__ Compared with an electron for which $n = 2$, an electron for which $n = 4$ has more
   - (a) spin.
   - (b) particle nature.
   - (c) energy.
   - (d) wave nature.

6. __c__ According to Bohr, which is the point in the figure below where electrons cannot reside?
   - (a) point A
   - (b) point B
   - (c) point C
   - (d) point D

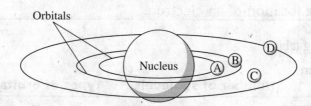

7. __c__ According to the quantum theory, point D in the above figure represents
   - (a) the fixed position of an electron.
   - (b) the farthest position from the nucleus that an electron can achieve.
   - (c) a position where an electron probably exists.
   - (d) a position where an electron cannot exist.

SECTION 2 continued

8. How did de Broglie conclude that electrons have a wave nature?

   Scientists knew that any wave confined to a space could have only certain frequencies. De Broglie suggested that electrons should be considered as waves confined to the space around an atomic nucleus; in this way, electron waves could exist only at specific frequencies. According to the relationship $E = h\nu$, these frequencies correspond to the specific quantized energies of the Bohr orbitals.

9. Identify each of the four quantum numbers and the properties to which they refer.

   The principal quantum number refers to the main energy level. The angular momentum quantum number refers to the shape of the orbital. The magnetic quantum number refers to the orientation of an orbital around the nucleus. The spin quantum number indicates the spin state of an electron in an orbital.

10. How did the Heisenberg uncertainty principle contribute to the idea that electrons occupy "clouds," or "orbitals"?

    The Heisenberg uncertainty principle states that it is impossible to determine simultaneously both the position and velocity of an electron (or any other particle). Because measuring the position of an electron actually changes its position, there is always a basic uncertainty in trying to locate an electron. Thus, the exact position of the electron cannot be found. An electron cloud or orbital represents the region that is the probable location of an electron.

11. Complete the following table:

Principal quantum number, n	Number of sublevels	Types of orbitals
1	1	s
2	2	s,p
3	3	s,p,d
4	4	s,p,d,f

CHAPTER 4 REVIEW
Arrangement of Electrons in Atoms

SECTION 3

SHORT ANSWER Answer the following questions in the space provided.

1. State the Pauli exclusion principle, and use it to explain why electrons in the same orbital must have opposite spin states.

   The Pauli exclusion principle states that no two electrons in an atom may have the same set of four quantum numbers. If both electrons in the same orbital had the same spin state, each electron would have the same four quantum numbers. If one electron has the opposite spin state, the fourth quantum number is different and the exclusion principle is obeyed.

2. Explain the conditions under which the following orbital notation for helium is possible:

   $\underline{\uparrow}$ $\underline{\uparrow}$
   $1s2s$

   This orbital notation is possible if the helium atom is in an excited state.

Write the ground-state electron configuration and orbital notation for each of the following atoms:

3. Phosphorus

   $1s^2 2s^2 2p^6 3s^2 3p^3$; $\underline{\uparrow\downarrow}$ $\underline{\uparrow\downarrow}$ $\underline{\uparrow\downarrow}$ $\underline{\uparrow\downarrow}$ $\underline{\uparrow\downarrow}$ $\underline{\uparrow\downarrow}$ $\underline{\uparrow}$ $\underline{\uparrow}$ $\underline{\uparrow}$
   $1s2s2p_x2p_y2p_z3s3p_x3p_y3p_z$

4. Nitrogen

   $1s^2 2s^2 2p^3$; $\underline{\uparrow\downarrow}$ $\underline{\uparrow\downarrow}$ $\underline{\uparrow}$ $\underline{\uparrow}$ $\underline{\uparrow}$
   $1s2s2p_x2p_y2p_z$

5. Potassium

   $1s^2 2s^2 2p^6 3s^2 3p^6 4s^1$; $\underline{\uparrow\downarrow}$ $\underline{\uparrow\downarrow}$ $\underline{\uparrow\downarrow}$ $\underline{\uparrow\downarrow}$ $\underline{\uparrow\downarrow}$ $\underline{\uparrow\downarrow}$ $\underline{\uparrow\downarrow}$ $\underline{\uparrow\downarrow}$ $\underline{\uparrow\downarrow}$ $\underline{\uparrow}$
   $1s2s2p_x2p_y2p_z3s3p_x3p_y3p_z4s$

MODERN CHEMISTRY

SECTION 3 continued

6. Aluminum

$1s^2 2s^2 2p^6 3s^2 3p^1$; $\underset{1s}{\uparrow\downarrow}$ $\underset{2s}{\uparrow\downarrow}$ $\underset{2p_x}{\uparrow\downarrow}$ $\underset{2p_y}{\uparrow\downarrow}$ $\underset{2p_z}{\uparrow\downarrow}$ $\underset{3s}{\uparrow\downarrow}$ $\underset{3p_x}{\uparrow}$ $\underset{3p_y}{}$ $\underset{3p_z}{}$

7. Argon

$1s^2 2s^2 2p^6 3s^2 3p^6$; $\underset{1s}{\uparrow\downarrow}$ $\underset{2s}{\uparrow\downarrow}$ $\underset{2p_x}{\uparrow\downarrow}$ $\underset{2p_y}{\uparrow\downarrow}$ $\underset{2p_z}{\uparrow\downarrow}$ $\underset{3s}{\uparrow\downarrow}$ $\underset{3p_x}{\uparrow\downarrow}$ $\underset{3p_y}{\uparrow\downarrow}$ $\underset{3p_z}{\uparrow\downarrow}$

8. Boron

$1s^2 2s^2 2p^1$; $\underset{1s}{\uparrow\downarrow}$ $\underset{2s}{\uparrow\downarrow}$ $\underset{2p_x}{\uparrow}$ $\underset{2p_y}{}$ $\underset{2p_z}{}$

9. Which guideline, Hund's rule or the Pauli exclusion principle, is violated in the following orbital diagrams?

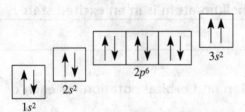

a. ___Pauli exclusion principle___

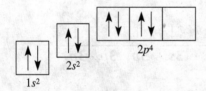

b. ___Hund's rule___

Name _____ Date _____ Class _____

CHAPTER 4 REVIEW
Arrangement of Electrons in Atoms

MIXED REVIEW

SHORT ANSWER Answer the following questions in the space provided.

1. Under what conditions is a photon emitted from an atom?
   A photon is emitted when an electron moves from a higher energy level to a lower energy level.

2. What do quantum numbers describe?
   Quantum numbers describe the energy level, orbital shape, orbital orientation, and spin state of an electron.

3. What is the relationship between the principal quantum number and the electron configuration?
   The principle quantum number, n, describes the energy level. For example, the electrons at $2p^6$ are at the energy level represented by $n = 2$.

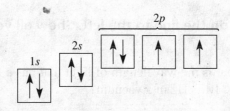

4. In what way does the figure above illustrate Hund's rule?
   The most stable arrangement of electrons is one with the maximum number of unpaired electrons.

5. In what way does the figure above illustrate the Pauli exclusion principle?
   No two electrons have the same set of four quantum numbers.

MIXED REVIEW continued

6. Elements of the fourth and higher main-energy levels do not seem to follow the normal sequence for filling orbitals. Why is this so?

   Electrons from the s orbital will sometimes be promoted to a higher energy level in order to form an electron configuration of lowest energy, which is therefore the most stable.

7. How do electrons create the colors in a line-emission spectrum?

   The colors are created by photons containing specific amounts of energy, released when an electron makes the transition from a higher energy level to a lower energy level.

8. Write the ground-state electron configuration of the following atoms:

   a. Carbon

   $1s^2 2s^2 2p^2$

   b. Potassium

   $1s^2 2s^2 2p^6 3s^2 3p^6 4s^1$

   c. Gallium

   $1s^2 2s^2 2p^6 3s^2 3p^6 3d^{10} 4s^2 4p^1$

   d. Copper

   $1s^2 2s^2 2p^6 3s^2 3p^6 3d^{10} 4s^1$

PROBLEMS Write the answer on the line to the left. Show all your work in the space provided.

9. _____1×10^{12} m_____ What is the wavelength of light that has a frequency of 3×10^{-4} Hz in a vacuum?

10. _____3.3×10^{-19} J_____ What is the energy of a photon that has a frequency of 5.0×10^{14} Hz?

CHAPTER 5 REVIEW

The Periodic Law

SECTION 1

SHORT ANSWER Answer the following questions in the space provided.

1. __c__ In the modern periodic table, elements are ordered

   (a) according to decreasing atomic mass.
   (b) according to Mendeleev's original design.
   (c) according to increasing atomic number.
   (d) based on when they were discovered.

2. __d__ Mendeleev noticed that certain similarities in the chemical properties of elements appeared at regular intervals when the elements were arranged in order of increasing

   (a) density.
   (b) reactivity.
   (c) atomic number.
   (d) atomic mass.

3. __b__ The modern periodic law states that

   (a) no two electrons with the same spin can be found in the same place in an atom.
   (b) the physical and chemical properties of an element are functions of its atomic number.
   (c) electrons exhibit properties of both particles and waves.
   (d) the chemical properties of elements can be grouped according to periodicity, but physical properties cannot.

4. __c__ The discovery of the noble gases changed Mendeleev's periodic table by adding a new

   (a) period.
   (b) series.
   (c) group.
   (d) level.

5. __d__ The most distinctive property of the noble gases is that they are

   (a) metallic.
   (b) radioactive.
   (c) metalloid.
   (d) largely unreactive.

6. __c__ Lithium, the first element in Group 1, has an atomic number of 3. The second element in this group has an atomic number of

   (a) 4.
   (b) 10.
   (c) 11.
   (d) 18.

7. An isotope of fluorine has a mass number of 19 and an atomic number of 9.

   __9__ a. How many protons are in this atom?

   __10__ b. How many neutrons are in this atom?

   __$^{19}_{9}F$__ c. What is the nuclear symbol of this fluorine atom, including its mass number and atomic number?

SECTION 1 continued

8. Samarium, Sm, is a member of the lanthanide series.

   __Pu, plutonium__ a. Identify the element just below samarium in the periodic table.

   __32 units__ b. By how many units do the atomic numbers of these two elements differ?

9. A certain isotope contains 53 protons, 78 neutrons, and 54 electrons.

   __53__ a. What is its atomic number?

   __131__ b. What is the mass number of this atom?

   __Iodine, I__ c. What is the name of this element?

   __F, Cl, Br, At__ d. Identify two other elements that are in the same group as this element.

10. In a modern periodic table, every element is a member of both a horizontal row and a vertical column. Which one is the group, and which one is the period?

    The group is the vertical column, and the period is the horizontal row.

11. Explain the distinction between atomic mass and atomic number of an element.

    The atomic number is the number of protons in an atom. The atomic mass is a weighted average of the masses of the naturally occurring isotopes of that element.

12. In the periodic table, the atomic number of I is greater than that of Te, but its atomic mass is less. This phenomenon also occurs with other neighboring elements in the periodic table. Name two of these pairs of elements. Refer to the periodic table if necessary.

    Co and Ni; Ar and K; Th and Pa; U and Np; Pu and Am; Sg and Bh. (The phenomenon occurs here because the mass of only the most stable isotope of each element is given.)

CHAPTER 5 REVIEW
The Periodic Law

SECTION 2

SHORT ANSWER Use this periodic table to answer the following questions in the space provided.

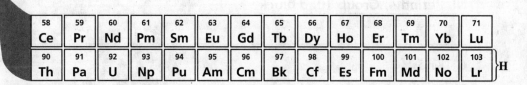

1. Identify the element and write the noble-gas notation for each of the following:

   a. the Group 14 element in Period 4
   Ge; $[Ar]3d^{10}4s^24p^2$

   b. the only metal in Group 15
   Bi; $[Xe]4f^{14}5d^{10}6s^26p^3$

   c. the transition metal with the smallest atomic mass
   Sc; $[Ar]3d^14s^2$

   d. the alkaline-earth metal with the largest atomic number
   Ra; $[Rn]7s^2$

MODERN CHEMISTRY   THE PERIODIC LAW **35**

SECTION 2 continued

2. On the periodic table given, several areas are labeled with letters A–H.

   _p_ block   a. Which block does **A** represent, *s, p, d,* or *f*?

   b. Identify the remaining labeled areas of the table, choosing from the following terms: *main-group elements, transition elements, lanthanides, actinides, alkali metals, alkaline-earth metals, halogens, noble gases.*

alkali metals	**B**
alkaline-earth metals	**C**
transition elements	**D**
main-group elements (also in B and C)	**E**
halogens	**F**
noble gases	**G**
actinides	**H**

3. Give the symbol, period, group, and block for the following:

   a. sulfur
   S, Period 3, Group 16, _p_ block

   b. nickel
   Ni, Period 4, Group 10, _d_ block

   c. $[Kr]5s^1$
   Rb, Period 5, Group 1, _s_ block

   d. $[Ar]3d^54s^1$
   Cr, Period 4, Group 6, _d_ block

4. There are 18 columns in the periodic table; each has a group number. Give the group numbers that make up each of the following blocks:

   1–2   a. *s* block

   13–18   b. *p* block

   3–12   c. *d* block

CHAPTER 5 REVIEW
The Periodic Law

SECTION 3

SHORT ANSWER Answer the following questions in the space provided.

1. __c__ When an electron is added to a neutral atom, energy is
   - (a) always absorbed.
   - (b) always released.
   - (c) either absorbed or released.
   - (d) neither absorbed nor released.

2. __d__ The energy required to remove an electron from a neutral atom is the atom's
   - (a) electron affinity.
   - (b) electron energy.
   - (c) electronegativity.
   - (d) neither absorbed nor released.

3. From left to right across a period on the periodic table,

   __negative__ a. electron affinity values tend to become more (negative or positive).

   __increase__ b. ionization energy values tend to (increase or decrease).

   __smaller__ c. atomic radii tend to become (larger or smaller).

4. __At__ a. Name the halogen with the least-negative electron affinity.

   __Li__ b. Name the alkali metal with the highest ionization energy.

   __Ar__ c. Name the element in Period 3 with the smallest atomic radius.

   __C__ d. Name the Group 14 element with the largest electronegativity.

5. Write the electron configuration of the following:

   a. Na
   $1s^2 2s^2 2p^6 3s^1$

   b. Na^+
   $1s^2 2s^2 2p^6$

   c. O
   $1s^2 2s^2 2p^4$

   d. O^{2-}
   $1s^2 2s^2 2p^6$

   e. Co^{2+}
   $1s^2 2s^2 2p^6 3s^2 3p^6 3d^7$

SECTION 3 continued

6. a. Compare the radius of a positive ion to the radius of its neutral atom.

The radius of a positive ion is smaller than the radius of its corresponding neutral atom.

b. Compare the radius of a negative ion to the radius of its neutral atom.

The radius of a negative ion is larger than the radius of its corresponding neutral atom.

7. a. Give the approximate positions and blocks where metals and nonmetals are found in the periodic table.

Metals are on the left side of the periodic table, mostly in the s, d, and f blocks.

Nonmetals are on the right side of the periodic table, all in the p block (except for hydrogen).

b. Of metals and nonmetals, which tend to form positive ions? Which tend to form negative ions?

Metals tend to form positive ions; nonmetals tend to form negative ions.

8. Table 3 on page 155 of the text lists successive ionization energies for several elements.

$3s^2$ **a.** Identify the electron that is removed in the first ionization energy of Mg.

$3s^1$ **b.** Identify the electron that is removed in the second ionization energy of Mg.

$2p^6$ **c.** Identify the electron that is removed in the third ionization energy of Mg.

d. Explain why the second ionization energy is higher than the first, the third is higher than the second, and so on.

As electrons are removed in successive ionizations, fewer electrons remain within the atom to shield the attractive force of the nucleus. Each electron removed from an ion experiences a stronger effective nuclear pull than the electron removed before it.

9. Explain the role of valence electrons in the formation of chemical compounds.

Valence electrons are the electrons most subject to the influence of nearby atoms or ions. They are the electrons available to be lost, gained, or shared in the formation of chemical compounds.

CHAPTER 5 REVIEW
The Periodic Law

MIXED REVIEW

SHORT ANSWER Answer the following questions in the space provided.

1. Consider the neutral atom with 53 protons and 74 neutrons to answer the following questions.

   __53__ a. What is its atomic number?

   __127__ b. What is its mass number?

   __atomic number__ c. Is the element's position in a modern periodic table determined by its atomic number or by its atomic mass?

2. Consider an element whose outermost electron configuration is $3d^{10}4s^24p^x$.

   __Period 4__ a. To which period does the element belong?

   __5__ b. If it is a halogen, what is the value of x?

   __True__ c. The group number will equal $(10 + 2 + x)$. True or False?

3. __p__ a. In which block are metalloids found, s, p, d, or f?

   __d__ b. In which block are the hardest, densest metals found, s, p, or d?

4. __fluorine, F__ a. Name the most chemically active halogen.

   __$1s^22s^22p^5$__ b. Write its electron configuration.

   __$1s^22s^22p^6$ for 1− ion__ c. Write the configuration of the most stable ion this element makes.

5. Refer only to the periodic table at the top of the review of Section 2 to answer the following questions on periodic trends.

   __In__ a. Which has the larger radius, Al or In?

   __Ca__ b. Which has the larger radius, Se or Ca?

   __Ca__ c. Which has a larger radius, Ca or Ca^{2+}?

   __nonmetals__ d. Which class has greater ionization energies, metals or nonmetals?

   __Cl__ e. Which has the greater ionization energy, As or Cl?

   __negative ion__ f. An element with a large negative electron affinity is most likely to form a (positive ion, negative ion, or neutral atom)?

MODERN CHEMISTRY THE PERIODIC LAW

MIXED REVIEW continued

____small____ g. In general, which has a stronger electron attraction, a large atom or a small atom?

____O____ h. Which has greater electronegativity, O or Se?

____O____ i. In the covalent bond between Se and O, to which atom is the electron pair more closely drawn?

____6____ j. How many valence electrons are there in a neutral atom of Se?

6. ____Ca^+ and Zn^{2+}____ Identify all of the following ions that do not have noble-gas stability.
K^+ S^{2-} Ca^+ I^- Al^{3+} Zn^{2+}

7. Use only the periodic table in the review of Section 2 to give the noble-gas notation of the following:

____$[Ar]3d^{10}4s^24p^5$____ a. Br

____$[Ar]3d^{10}4s^24p^6$____ b. Br^-

____$[Kr]4d^{10}5s^25p^1$____ c. the element in Group 13, Period 5

____$[Xe]4f^15d^16s^2$____ d. the lanthanide with the smallest atomic number

8. Use electron configuration and position in the periodic table to describe the chemical properties of calcium and oxygen.

Calcium is a Group 2 alkaline-earth metal with $[Ar]4s^2$ configuration. It forms a stable 2+ ion, has relatively low ionization energy, and forms salt-like ionic compounds. Oxygen, with $[He]2s^22p^4$ configuration, is a typical Group 16 nonmetal, making a stable 2− ion; it has high electronegativity and ionization energy and quite negative electron affinity.

9. Copper's electron configuration might be predicted to be $3d^94s^2$. But in fact, its configuration is $3d^{10}4s^1$. The two elements below copper in Group 11 behave similarly. (Confirm this in the periodic table in **Figure 6** on pages 140–141 of the text.)

____$3d^{10}4s^1$____ a. Which configuration for copper is apparently more stable?

____Yes____ b. Is the d sublevel completed in the atoms of these three elements?

____True____ c. Every element in Period 4 has four levels of electrons established. True or False?

Name _____ Date _____ Class _____

CHAPTER 6 REVIEW
Chemical Bonding

SECTION 1

SHORT ANSWER Answer the following questions in the space provided.

1. __a__ A chemical bond between atoms results from the attraction between the valence electrons and ____ of different atoms.
   - (a) nuclei
   - (b) inner electrons
   - (c) isotopes
   - (d) Lewis structures

2. __b__ A covalent bond consists of
   - (a) a shared electron.
   - (b) a shared electron pair.
   - (c) two different ions.
   - (d) an octet of electrons.

3. __a__ If two covalently bonded atoms are identical, the bond is identified as
   - (a) nonpolar covalent.
   - (b) polar covalent.
   - (c) ionic.
   - (d) dipolar.

4. __b__ A covalent bond in which there is an unequal attraction for the shared electrons is
   - (a) nonpolar.
   - (b) polar.
   - (c) ionic.
   - (d) dipolar.

5. __c__ Atoms with a strong attraction for electrons they share with another atom exhibit
   - (a) zero electronegativity.
   - (b) low electronegativity.
   - (c) high electronegativity.
   - (d) Lewis electronegativity.

6. __c__ Bonds that possess between 5% and 50% ionic character are considered to be
   - (a) ionic.
   - (b) pure covalent.
   - (c) polar covalent.
   - (d) nonpolar covalent.

7. __a__ The greater the electronegativity difference between two atoms bonded together, the greater the bond's percentage of
   - (a) ionic character.
   - (b) nonpolar character.
   - (c) metallic character.
   - (d) electron sharing.

8. The electrons involved in the formation of a chemical bond are called __valence electrons__.

9. A chemical bond that results from the electrostatic attraction between positive and negative ions is called a(n) __ionic bond__.

MODERN CHEMISTRY   CHEMICAL BONDING **41**
Copyright © by Holt, Rinehart and Winston. All rights reserved.

Name _____ Date _____ Class _____

SECTION 1 continued

10. If electrons involved in bonding spend most of the time closer to one atom rather than the other, the bond is _____ polar covalent _____.

11. If a bond's character is more than 50% ionic, then the bond is called a(n) _____ ionic bond _____.

12. A bond's character is more than 50% ionic if the electronegativity difference between the two atoms is greater than _____ 1.7 _____.

13. Write the formula for an example of each of the following compounds:

    Answers will vary.
    - _____ H_2 _____ a. nonpolar covalent compound
    - _____ HCl _____ b. polar covalent compound
    - _____ NaCl _____ c. ionic compound

14. Describe how a covalent bond holds two atoms together.
    A pair of electrons is attracted to both nuclei of the two atoms bonded together.

15. What property of the two atoms in a covalent bond determines whether or not the bond will be polar?
    electronegativity

16. How can electronegativity be used to distinguish between an ionic bond and a covalent bond?
    The difference between the electronegativity of the two atoms in a bond will determine whether the bond is ionic or covalent. If the difference in electronegativity is greater than 1.7, the bond is considered ionic.

17. Describe the electron distribution in a polar-covalent bond and its effect on the partial charges of the compound.
    The electron density is greater around the more electronegative atom, giving that part of the compound a partial negative charge. The other part of the compound has an equal partial positive charge.

CHEMICAL BONDING   MODERN CHEMISTRY

CHAPTER 6 REVIEW
Chemical Bonding

SECTION 2

SHORT ANSWER Answer the following questions in the space provided.

1. Use the concept of potential energy to describe how a covalent bond forms between two atoms.

   As the atoms involved in the formation of a covalent bond approach each other, the electron-proton attraction is stronger than the electron-electron and proton-proton repulsions. The atoms are drawn to each other and their potential energy decreases. Eventually, a distance is reached at which the repulsions between the like charges equals the attraction of the opposite charges. At this point, potential energy is at a minimum and a stable molecule forms.

2. Name two elements that form compounds that can be exceptions to the octet rule.

   Choose from hydrogen, boron, beryllium, phosphorus, sulfur, and xenon.

3. Explain why resonance structures are used instead of Lewis structures to correctly model certain molecules.

   Resonance structures show that one Lewis structure cannot correctly represent the location of electrons in a bond. Resonance structures show delocalized electrons, while Lewis structures depict electrons in a definite location.

4. Bond energy is related to bond length. Use the data in the tables below to arrange the bonds listed in order of increasing bond length, from shortest bond to longest.

   a.

Bond	Bond energy (kJ/mol)
H—F	569
H—I	299
H—Cl	432
H—Br	366

   H—F, H—Cl, H—Br, H—I

MODERN CHEMISTRY CHEMICAL BONDING 43

SECTION 2 continued

b.

Bond	Bond energy (kJ/mol)
C—C	346
C≡C	835
C=C	612

C≡C, C=C, C—C

---

5. Draw Lewis structures to represent each of the following formulas:

a. NH_3

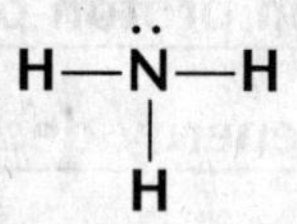

```
H—N̈—H
    |
    H
```

b. H_2O

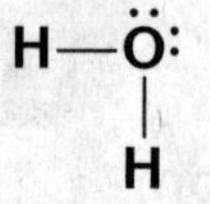

```
H—Ö:
|
H
```

c. CH_4

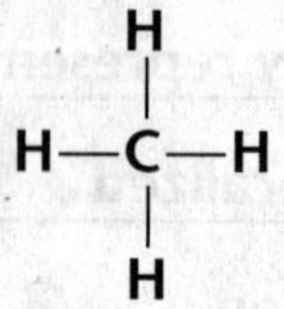

d. C_2H_2

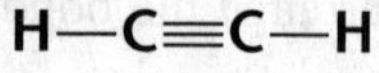

H—C≡C—H

e. CH_2O

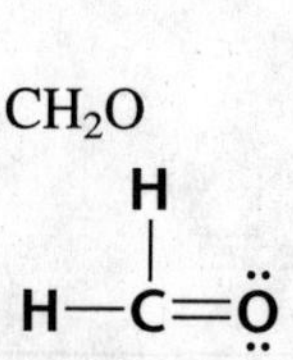

```
    H
    |
H—C=Ö
```

CHAPTER 6 REVIEW
Chemical Bonding

SECTION 3

SHORT ANSWER Answer the following questions in the space provided.

1. __a__ The notation for sodium chloride, NaCl, stands for one
   - (a) formula unit.
   - (b) molecule.
   - (c) crystal.
   - (d) atom.

2. __d__ In a crystal of an ionic compound, each cation is surrounded by a number of
   - (a) molecules.
   - (b) positive ions.
   - (c) dipoles.
   - (d) negative ions.

3. __b__ Compared with the neutral atoms involved in the formation of an ionic compound, the crystal lattice that results is
   - (a) higher in potential energy.
   - (b) lower in potential energy.
   - (c) equal in potential energy.
   - (d) unstable.

4. __b__ The lattice energy of compound A is greater in magnitude than that of compound B. What can be concluded from this fact?
   - (a) Compound A is not an ionic compound.
   - (b) It will be more difficult to break the bonds in compound A than those in compound B.
   - (c) Compound B has larger crystals than compound A.
   - (d) Compound A has larger crystals than compound B.

5. __b__ The forces of attraction between molecules in a molecular compound are generally
   - (a) stronger than the attractive forces among formula units in ionic bonding.
   - (b) weaker than the attractive forces among formula units in ionic bonding.
   - (c) approximately equal to the attractive forces among formula units in ionic bonding.
   - (d) equal to zero.

6. Describe the force that holds two ions together in an ionic bond.

   The force of attraction between unlike charges holds a negative ion and a positive ion together in an ionic bond.

7. What type of energy best represents the strength of an ionic bond?

   lattice energy

SECTION 3 continued

8. What types of bonds are present in an ionic compound that contains a polyatomic ion?

The atoms in a polyatomic ion are held together with covalent bonds, but

polyatomic ions combine with ions of opposite charge to form ionic compounds.

9. Arrange the ionic bonds in the table below in order of increasing strength from weakest to strongest.

Ionic bond	Lattice energy (kJ/mol)
NaCl	−787
CaO	−3384
KCl	−715
MgO	−3760
LiCl	−861

KCl, NaCl, LiCl, CaO, MgO

10. Draw Lewis structures for the following polyatomic ions:

a. NH_4^+

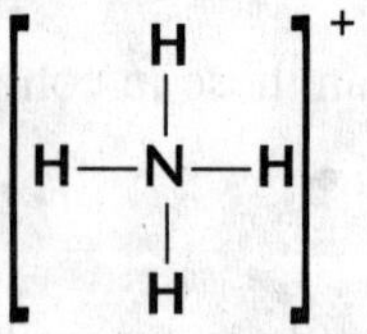

$$\left[\begin{array}{c} H \\ | \\ H-N-H \\ | \\ H \end{array} \right]^+$$

b. SO_4^{2-}

$$\left[\begin{array}{c} \ddot{O}: \\ | \\ :\ddot{O}-S-\ddot{O}: \\ | \\ :\ddot{O}: \end{array} \right]^{2-}$$

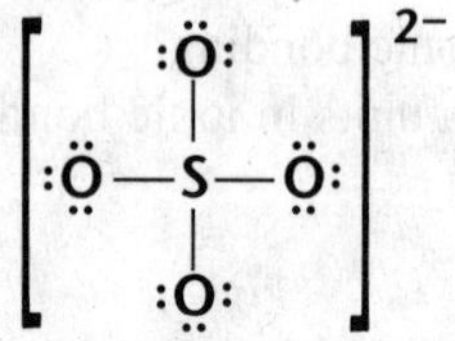

11. Draw the two resonance structures for the nitrite anion, NO_2^-.

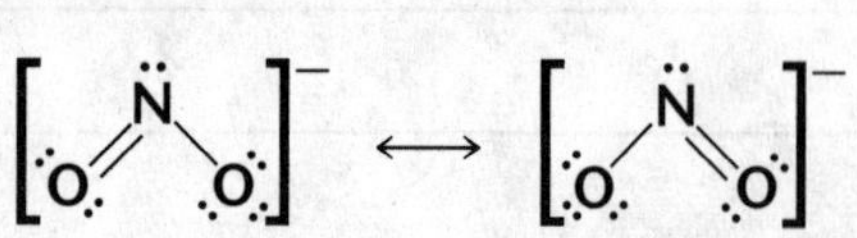

$$\left[:\ddot{O}-\ddot{N}=\ddot{O}: \right]^- \longleftrightarrow \left[:\ddot{O}=\ddot{N}-\ddot{O}: \right]^-$$

CHAPTER 6 REVIEW
Chemical Bonding

SECTION 4

SHORT ANSWER Answer the following questions in the space provided.

1. __b__ In metals, the valence electrons are considered to be
   - (a) attached to particular positive ions.
   - (b) shared by all surrounding atoms.
   - (c) immobile.
   - (d) involved in covalent bonds.

2. __a__ The fact that metals are malleable and ionic crystals are brittle is best explained in terms of their
   - (a) chemical bonds.
   - (b) London forces.
   - (c) enthalpies of vaporization.
   - (d) polarity.

3. __d__ As light strikes the surface of a metal, the electrons in the electron sea
   - (a) allow the light to pass through.
   - (b) become attached to particular positive ions.
   - (c) fall to lower energy levels.
   - (d) absorb and re-emit the light.

4. __d__ Mobile electrons in the metallic bond are responsible for
   - (a) luster.
   - (b) thermal conductivity.
   - (c) electrical conductivity.
   - (d) All of the above.

5. __a__ In general, the strength of the metallic bond ____ moving from left to right on any row of the periodic table.
   - (a) increases
   - (b) decreases
   - (c) remains the same
   - (d) varies

6. __c__ When a metal is drawn into a wire, the metallic bonds
   - (a) break easily.
   - (b) break with difficulty.
   - (c) do not break.
   - (d) become ionic bonds.

7. Use the concept of electron configurations to explain why the number of valence electrons in metals tends to be less than the number in most nonmetals.

   Most metals have their outer electrons in *s* orbitals, while nonmetals have their

   outer electrons in *p* orbitals.

SECTION 4 continued

8. How does the behavior of electrons in metals contribute to the metal's ability to conduct electricity and heat?

The mobility of electrons in a network of metal atoms contributes to the

metal's ability to conduct electricity and heat.

9. What is the relationship between the enthalpy of vaporization of a metal and the strength of the bonds that hold the metal together?

The amount of energy required to vaporize a metal is a measure of the strength

of the bonds that hold the metal together. The greater a metal's enthalpy of

vaporization, the stronger the metallic bond.

10. Draw two diagrams of a metallic bond. In the first diagram, draw a weak metallic bond; in the second, show a metallic bond that would be stronger. Be sure to include nuclear charge and number of electrons in your illustrations.

a.

weak bond

b.

strong bond

Note: In the strong bond, the charge on the nucleus and the number of electrons must be greater than in the weak bond.

11. Complete the following table:

	Metals	Ionic Compounds
Components	atoms	ions
Overall charge	neutral	neutral
Conductive in the solid state	yes	no
Melting point	low to high	high
Hardness	soft to hard	hard
Malleable	yes	no
Ductile	yes	no

CHAPTER 6 REVIEW
Chemical Bonding

SECTION 5

SHORT ANSWER Answer the following questions in the space provided.

1. Identify the major assumption of the VSEPR theory, which is used to predict the shape of atoms.
   Pairs of valence electrons repel one another.

2. In water, two hydrogen atoms are bonded to one oxygen atom. Why isn't water a linear molecule?
   The electron pairs that are not involved in bonding also take up space, creating a tetrahedron of electron pairs and making the water molecule angular or bent.

3. What orbitals combine together to form sp^3 hybrid orbitals around a carbon atom?
   the s orbital and all three p orbitals from the second energy level

4. What two factors determine whether or not a molecule is polar?
   electronegativity difference and molecular geometry or unshared electron pairs

5. Arrange the following types of attractions in order of increasing strength, with 1 being the weakest and 4 the strongest.

   __3__ hydrogen bonding

   __4__ ionic

   __2__ dipole-dipole

   __1__ London dispersion

6. How are dipole-dipole attractions, London dispersion forces, and hydrogen bonding similar?
   They are all forces of attraction between molecules. In all cases there is an attraction between the slightly negatively-charged portion of one molecule and the slightly positively charged portion of another molecule.

SECTION 5 continued

7. Complete the following table:

Formula	Lewis structure	Geometry	Polar
H₂S	(H-S-H with lone pairs on S)	bent	yes
CCl₄	(C bonded to four Cl atoms)	tetrahedral	no
BF₃	(B bonded to three F atoms)	trigonal planar	no
H₂O	(H-O-H with lone pairs on O)	bent	yes
PCl₅	(P bonded to five Cl atoms)	trigonal bipyramidal	no
BeF₂	:F—Be—F:	linear	no
SF₆	(S bonded to six F atoms)	octahedral	no

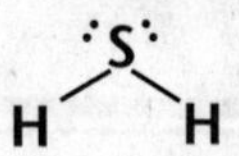

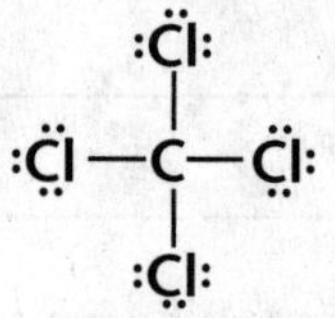

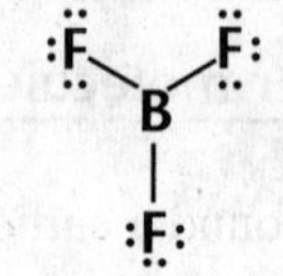

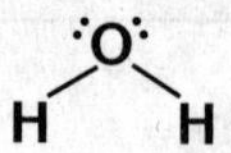

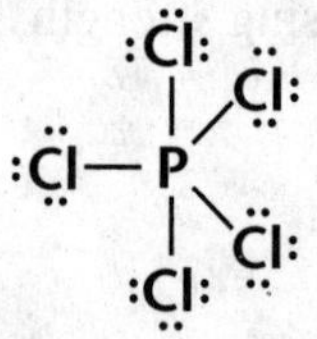

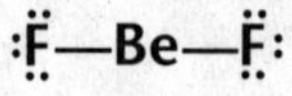

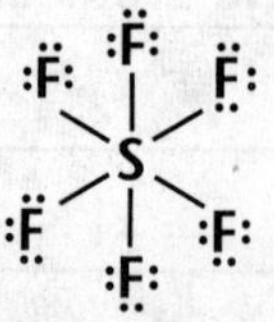

50 CHEMICAL BONDING

Name _____ Date _____ Class _____

CHAPTER 6 REVIEW
Chemical Bonding

MIXED REVIEW

SHORT ANSWER Answer the following questions in the space provided.

1. Name the type of energy that is a measure of strength for each of the following types of bonds:

    __lattice energy__ a. ionic bond

    __bond energy__ b. covalent bond

    __enthalpy of vaporization__ c. metallic bond

2. Use the electronegativity values shown in **Figure 20,** on page 161 of the text, to determine whether each of the following bonds is nonpolar covalent, polar covalent, or ionic.

    __ionic__ a. H—F   __nonpolar covalent__ d. H—H

    __ionic__ b. Na—Cl   __polar covalent__ e. H—C

    __polar covalent__ c. H—O __polar covalent__ f. H—N

3. How is a hydrogen bond different from an ionic or covalent bond?

    A hydrogen bond is a dipole-dipole attraction between a partially positive hydrogen

    atom and the unshared electron pair of a strongly electronegative atom such as O,

    N, or F. Unlike ionic or covalent bonds, in which electrons are given up or shared,

    the hydrogen bond is a weaker attraction. Hydrogen bonds are generally

    intermolecular, while ionic and covalent bonds occur between ions or atoms

    respectively.

4. H_2S and H_2O have similar structures and their central atoms belong to the same group. Yet H_2S is a gas at room temperature and H_2O is a liquid. Use bonding principles to explain why this is.

    Oxygen has higher electronegativity than sulfur, which creates a highly polar bond.

    Increased polarity in H_2O bonds means a stronger intermolecular attraction, making

    water a liquid at room temperature. Hydrogen bonding exists between water

    molecules, but not between hydrogen sulfide molecules.

MIXED REVIEW continued

5. In what way is a polar-covalent bond similar to an ionic bond?
There is a difference between the electronegativities of the two atoms in both types of bonds that results in electrons being more closely associated with the more electronegative atom.

6. Draw a Lewis structure for each of the following formulas. Determine whether the molecule is polar or nonpolar.

_____polar_____ **a.** H_2S

_____polar_____ **b.** $COCl_2$

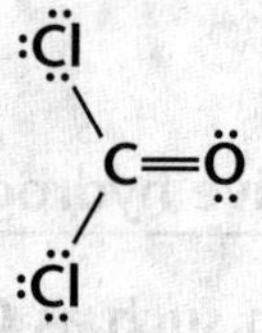

:Cl:
 |
 C=Ö
 |
:Cl:

_____polar_____ **c.** PCl_3

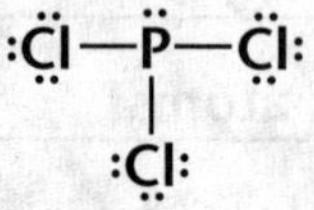

:Cl—P̈—Cl:
       |
     :Cl:

_____polar_____ **d.** CH_2O

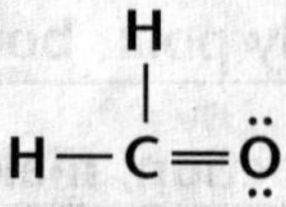

  H
  |
H—C=Ö

CHAPTER 7 REVIEW
Chemical Formulas and Chemical Compounds

SECTION 1

SHORT ANSWER Answer the following questions in the space provided.

1. __c__ In a Stock system name such as iron(III) sulfate, the Roman numeral tells us
    (a) how many atoms of Fe are in one formula unit.
    (b) how many sulfate ions can be attached to the iron atom.
    (c) the charge on each Fe ion.
    (d) the total positive charge of the formula unit.

2. __c__ Changing a subscript in a correctly written chemical formula
    (a) changes the number of moles represented by the formula.
    (b) changes the charges on the other ions in the compound.
    (c) changes the formula so that it no longer represents the compound it previously represented.
    (d) has no effect on the formula.

3. The explosive TNT has the molecular formula $C_7H_5(NO_2)_3$.

    __4 elements__ a. How many elements make up this compound?

    __6 oxygen atoms__ b. How many oxygen atoms are present in one molecule of $C_7H_5(NO_2)_3$?

    __21 atoms__ c. How many atoms in total are present in one molecule of $C_7H_5(NO_2)_3$?

    __4.2×10^{24} atoms__ d. How many atoms are present in a sample of 2.0×10^{23} molecules of $C_7H_5(NO_2)_3$?

4. How many atoms are present in each of these formula units?

    __11 atoms__ a. $Ca(HCO_3)_2$

    __45 atoms__ b. $C_{12}H_{22}O_{11}$

    __10 atoms__ c. $Fe(ClO_2)_3$

    __9 atoms__ d. $Fe(ClO_3)_2$

5. __N_2O_5__ a. What is the formula for the compound dinitrogen pentoxide?

    __iron(II) oxide__ b. What is the Stock system name for the compound FeO?

    __H_2SO_3__ c. What is the formula for sulfurous acid?

    __phosphoric acid__ d. What is the name for the acid H_3PO_4?

MODERN CHEMISTRY — CHEMICAL FORMULAS AND CHEMICAL COMPOUNDS

SECTION 1 continued

6. Some binary compounds are ionic, others are covalent. The type of bond favored partially depends on the position of the elements in the periodic table. Label each of these claims as True or False; if False, specify the nature of the error.

   a. Covalently bonded binary molecular compounds are typically composed of nonmetals.

   True

   b. Binary ionic compounds are composed of metals and nonmetals, typically from opposite sides of the periodic table.

   True

7. Refer to **Table 2** on page 226 of the text and **Table 5** on page 230 of the text for examples of names and formulas for polyatomic ions and acids.

   a. Derive a generalization for determining whether an acid name will end in the suffix -ic or -ous.

   In general, if the anion name ends in -ate, the corresponding acid name will end in a suffix of -ic. In general, if the anion name ends in -ite, the corresponding acid name will end in a suffix of -ous.

   b. Derive a generalization for determining whether an acid name will begin with the prefix hydro- or not.

   In general, if the anion name ends in -ide, the corresponding acid name will end in a suffix of -ic and begin with a prefix of hydro-. The prefix hydro- is never used for anions ending in -ate or -ite.

8. Fill in the blanks in the table below.

Compound name	Formula
Aluminum sulfide	Al_2S_3
Cesium carbonate	Cs_2CO_3
Lead(II) chloride	$PbCl_2$
Ammonium phosphate	$(NH_4)_3PO_4$
Hydroiodic acid	HI

Name _____ Date _____ Class _____

CHAPTER 7 REVIEW
Chemical Formulas and Chemical Compounds

SECTION 2

SHORT ANSWER Answer the following questions in the space provided.

1. Assign the oxidation number to the specified element in each of the following examples:

    __+4__ a. S in H_2SO_3

    __+6__ b. S in $MgSO_4$

    __−2__ c. S in K_2S

    __+1__ d. Cu in Cu_2S

    __+6__ e. Cr in Na_2CrO_4

    __+5__ f. N in HNO_3

    __+4__ g. C in $(HCO_3)^-$

    __−3__ h. N in $(NH_4)^+$

2. __SCl_2__ a. What is the formula for the compound sulfur(II) chloride?

    __nitrogen(IV) oxide__ b. What is the Stock system name for NO_2?

3. __fluorine__ a. Use electronegativity values to determine the one element that always has a negative oxidation number when it appears in any binary compound.

    __0; F_2__ b. What is the oxidation number and formula for the element described in part **a** when it exists as a pure element?

4. Tin has possible oxidation numbers of +2 and +4 and forms two known oxides. One of them has the formula SnO_2.

    __tin(IV) oxide__ a. Give the Stock system name for SnO_2.

    __SnO__ b. Give the formula for the other oxide of tin.

5. Scientists think that two separate reactions contribute to the depletion of the ozone, O_3, layer. The first reaction involves oxides of nitrogen. The second involves free chlorine atoms. The equations that represent the reactions follow. When a compound is not stated as a formula, write the correct formula in the blank beside its name.

    a. __NO__ (nitrogen monoxide) + O_3 → __NO_2__ (nitrogen dioxide) + O_2

SECTION 2 continued

b. Cl + O$_3$ → __ClO__ (chlorine monoxide) + O$_2$

6. Consider the covalent compound dinitrogen trioxide when answering the following:

   __N$_2$O$_3$__ a. What is the formula for dinitrogen trioxide?

   __+3__ b. What is the oxidation number assigned to each nitrogen atom in this compound? Explain your answer.

   The three oxygen atoms have oxidation states of −6 total, and because the algebraic sum of the oxidation states in a neutral compound must be zero, the two nitrogen atoms must have oxidation states of +6 total, therefore +3 each.

   __nitrogen(III) oxide__ c. Give the Stock name for dinitrogen trioxide.

7. The oxidation numbers assigned to the atoms in some organic compounds have unexpected values. Assign oxidation numbers to each atom in the following compounds: (Note: Some oxidation numbers may not be whole numbers.)

   a. CO$_2$
   Carbon is +4 and each oxygen is −2.

   b. CH$_4$ (methane)
   Carbon is −4 and each hydrogen is +1.

   c. C$_6$H$_{12}$O$_6$ (glucose)
   Each carbon is 0, each hydrogen is +1, and each oxygen is −2.

   d. C$_3$H$_8$ (propane gas)
   Each carbon is −8/3 and each hydrogen is +1.

8. Assign oxidation numbers to each element in the compounds found in the following situations:

   a. Rust, Fe$_2$O$_3$, forms on an old nail.
   Each iron is +3 and each oxygen is −2.

   b. Nitrogen dioxide, NO$_2$, pollutes the air as a component of smog.
   Nitrogen is +4 and each oxygen is −2.

   c. Chromium dioxide, CrO$_2$, is used to make recording tapes.
   Chromium is +4 and each oxygen is −2.

CHAPTER 7 REVIEW
Chemical Formulas and Chemical Compounds

SECTION 3

SHORT ANSWER Answer the following questions in the space provided.

1. Label each of the following statements as True or False:

   __True__ a. If the formula mass of one molecule is x amu, the molar mass is x g/mol.

   __False__ b. Samples of equal numbers of moles of two different chemicals must have equal masses as well.

   __True__ c. Samples of equal numbers of moles of two different molecular compounds must have equal numbers of molecules as well.

2. How many moles of each element are present in a 10.0 mol sample of $Ca(NO_3)_2$?

   10 mol of calcium, 20 mol of nitrogen, 60 mol of oxygen

PROBLEMS Write the answer on the line to the left. Show all your work in the space provided.

3. Consider a sample of 10.0 g of the gaseous hydrocarbon C_3H_4 to answer the following questions.

   __0.250 mol__ a. How many moles are present in this sample?

   __1.50×10^{23} molecules__ b. How many molecules are present in the C_3H_4 sample?

   __4.51×10^{23} carbon atoms__ c. How many carbon atoms are present in this sample?

SECTION 3 continued

___10.1%___ **d.** What is the percentage composition of hydrogen in the sample?

4. One source of aluminum metal is alumina, Al_2O_3.

___52.9%___ **a.** Determine the percentage composition of Al in alumina.

___2100 lb___ **b.** How many pounds of aluminum can be extracted from 2.0 tons of alumina?

5. Compound A has a molar mass of 20 g/mol, and compound B has a molar mass of 30 g/mol.

___20 g___ **a.** What is the mass of 1.0 mol of compound A, in grams?

___0.17 mol___ **b.** How many moles are present in 5.0 g of compound B?

___4.0 mol___ **c.** How many moles of compound B are needed to have the same mass as 6.0 mol of compound A?

CHAPTER 7 REVIEW
Chemical Formulas and Chemical Compounds

SECTION 4

SHORT ANSWER Answer the following questions in the space provided.

1. Write empirical formulas to match the following molecular formulas:

   __CH_3O_2__ a. $C_2H_6O_4$

   __N_2O_5__ b. N_2O_5

   __HgCl__ c. Hg_2Cl_2

   __CH_2__ d. C_6H_{12}

2. __C_4H_8__ A certain hydrocarbon has an empirical formula of CH_2 and a molar mass of 56.12 g/mol. What is its molecular formula?

3. A certain ionic compound is found to contain 0.012 mol of sodium, 0.012 mol of sulfur, and 0.018 mol of oxygen.

   __$Na_2S_2O_3$__ a. What is its empirical formula?

   __neither__ b. Is this compound a sulfate, sulfite, or neither?

PROBLEMS Write the answer on the line to the left. Show all your work in the space provided.

4. Water of hydration was discussed in **Sample Problem K** on pages 243–244 of the text. Strong heating will drive off the water as a vapor in hydrated copper(II) sulfate. Use the data table below to answer the following:

Mass of the empty crucible	4.00 g
Mass of the crucible plus hydrate sample	4.50 g
Mass of the system after heating	4.32 g
Mass of the system after a second heating	4.32 g

   __36%__ a. Determine the mass percentage of water in the original sample.

SECTION 4 continued

_____5_____ b. The compound has the formula $CuSO_4 \cdot xH_2O$. Determine the value of x.

c. What might be the purpose of the second heating?

The second heating is to ensure that all the water in the sample has been driven off. If the mass is less after the second heating, water was still present after the first heating.

5. Gas X is found to be 24.0% carbon and 76.0% fluorine by mass.

_____CF_2_____ a. Determine the empirical formula of gas X.

_____C_4F_8_____ b. Given that the molar mass of gas X is 200.04 g/mol, determine its molecular formula.

6. A compound is found to contain 43.2% copper, 24.1% chlorine, and 32.7% oxygen by mass.

_____$CuClO_3$_____ a. Determine its empirical formula.

b. What is the correct Stock system name of the compound in part **a**?
copper(I) chlorate

CHAPTER 7 REVIEW
Chemical Formulas and Chemical Compounds

MIXED REVIEW

SHORT ANSWER Answer the following questions in the space provided.

1. Write formulas for the following compounds:

   _____$CuCO_3$_____ a. copper(II) carbonate

   _____Na_2SO_3_____ b. sodium sulfite

   _____$(NH_4)_3PO_4$_____ c. ammonium phosphate

   _____SnS_2_____ d. tin(IV) sulfide

   _____HNO_2_____ e. nitrous acid

2. Write the Stock system names for the following compounds:

   _____magnesium perchlorate_____ a. $Mg(ClO_4)_2$

   _____iron(II) nitrate_____ b. $Fe(NO_3)_2$

   _____iron(III) nitrite_____ c. $Fe(NO_2)_3$

   _____cobalt(II) oxide_____ d. CoO

   _____nitrogen(V) oxide_____ e. dinitrogen pentoxide

3. _____13 atoms_____ a. How many atoms are represented by the formula $Ca(HSO_4)_2$?

   _____4.0 mol_____ b. How many moles of oxygen atoms are in a 0.50 mol sample of this compound?

   _____+6_____ c. Assign the oxidation number to sulfur in the HSO_4^- anion.

4. Assign the oxidation number to the element specified in each of the following:

   _____+1_____ a. hydrogen in H_2O_2

   _____−1_____ b. hydrogen in MgH_2

   _____0_____ c. sulfur in S_8

   _____+4_____ d. carbon in $(CO_3)^{2-}$

   _____+6_____ e. chromium in $Na_2Cr_2O_7$

   _____+4_____ f. nitrogen in NO_2

MODERN CHEMISTRY   CHEMICAL FORMULAS AND CHEMICAL COMPOUNDS

Name _____ Date _____ Class _____

MIXED REVIEW continued

PROBLEMS Write the answer on the line to the left. Show all your work in the space provided.

5. ____c, b, d, a____ Following are samples of four different compounds. Arrange them in order of increasing mass, from smallest to largest.

    a. 25 g of oxygen gas   **c.** 3×10^{23} molecules of C_2H_6
    b. 1.00 mol of H_2O **d.** 2×10^{23} molecules of $C_2H_6O_2$

6. ____NaOH____   **a.** What is the formula for sodium hydroxide?

   ____40.00 g/mol____   **b.** What is the formula mass of sodium hydroxide?

   ____10. g____   **c.** What is the mass in grams of 0.25 mol of sodium hydroxide?

7. ____80% C, 20% H____ What is the percentage composition of ethane gas, C_2H_6, to the nearest whole number?

8. ____$C_5H_{10}O_5$____ Ribose is an important sugar (part of RNA), with a molar mass of 150.15 g/mol. If its empirical formula is CH_2O, what is its molecular formula?

MIXED REVIEW continued

9. Butane gas, C_4H_{10}, is often used as a fuel.

   __174 g__ a. What is the mass in grams of 3.00 mol of butane?

   __1.81×10^{24} molecules__ b. How many molecules are present in that 3.00 mol sample?

   __C_2H_5__ c. What is the empirical formula of the gas?

10. __$C_{10}H_8$__ Naphthalene is a soft covalent solid that is often used in mothballs. Its molar mass is 128.18 g/mol and it contains 93.75% carbon and 6.25% hydrogen. Determine the molecular formula of napthalene from this information.

11. Nicotine has the formula $C_xH_yN_z$. To determine its composition, a sample is burned in excess oxygen, producing the following results:

    1.0 mol of CO_2
    0.70 mol of H_2O
    0.20 mol of NO_2

    Assume that all the atoms in nicotine are present as products.

    __1.0 mol__ a. Determine the number of moles of carbon present in the products of this combustion.

MIXED REVIEW continued

__1.40 mol__   b. Determine the number of moles of hydrogen present in the combustion products.

__0.20 mol__   c. Determine the number of moles of nitrogen present in the combustion products.

__C_5H_7N__   d. Determine the empirical formula of nicotine based on your calculations.

__162 g/mol__   e. In a separate experiment, the molar mass of nicotine is found to be somewhere between 150 and 180 g/mol. Calculate the molar mass of nicotine to the nearest gram.

12. When $MgCO_3(s)$ is strongly heated, it produces solid MgO as gaseous CO_2 is driven off.

__52.2%__   a. What is the percentage loss in mass as this reaction occurs?

__Mg is +2, C is +4, and O is −2__   b. Assign the oxidation number to each atom in $MgCO_3$.

__No__   c. Does the oxidation number of carbon change upon the formation of CO_2?

Name _____ Date _____ Class _____

CHAPTER 8 REVIEW
Chemical Equations and Reactions

SECTION 1

SHORT ANSWER Answer the following questions in the space provided.

1. Match the symbol on the left with its appropriate description on the right.

    __d__ $\xrightarrow{\Delta}$ (a) A precipitate forms.

    __a__ ↓ (b) A gas forms.

    __b__ ↑ (c) A reversible reaction occurs.

    __f__ (l) (d) Heat is applied to the reactants.

    __e__ (aq) (e) A chemical is dissolved in water.

    __c__ ⇌ (f) A chemical is in the liquid state.

2. Finish balancing the following equation:

    $3Fe_3O_4 + \underline{}8\underline{} Al \rightarrow \underline{}4\underline{} Al_2O_3 + \underline{}9\underline{} Fe$

3. In each of the following formulas, write the total number of atoms present.

    __12 atoms__ a. $4SO_2$

    __16 atoms__ b. $8O_2$

    __51 atoms__ c. $3Al_2(SO_4)_3$

    __3×10^{24} atoms__ d. 6×10^{23} HNO_3

4. Convert the following word equation into a balanced chemical equation:
   aluminum metal + copper(II) fluoride → aluminum fluoride + copper metal

   $2Al(s) + 3CuF_2(aq) \rightarrow 2AlF_3(aq) + 3Cu(s)$

5. One way to test the salinity of a water sample is to add a few drops of silver nitrate solution with a known concentration. As the solutions of sodium chloride and silver nitrate mix, a precipitate of silver chloride forms, and sodium nitrate is left in solution. Translate these sentences into a balanced chemical equation.

   $NaCl(aq) + AgNO_3(aq) \rightarrow AgCl(s) + NaNO_3(aq)$

6. a. Balance the following equation: $NaHCO_3(s) \xrightarrow{\Delta} Na_2CO_3(s) + H_2O(g) + CO_2(g)$

   $2NaHCO_3(s) \xrightarrow{\Delta} Na_2CO_3(s) + H_2O(g) + CO_2(g)$

SECTION 1 continued

b. Translate the chemical equation in part **a** into a sentence.

When solid sodium hydrogen carbonate (bicarbonate) is heated, it decomposes into solid sodium carbonate while releasing carbon dioxide gas and water vapor.

7. The poisonous gas hydrogen sulfide, H_2S, can be neutralized with a base such as sodium hydroxide, NaOH. The unbalanced equation for this reaction follows:

$$NaOH(aq) + H_2S(g) \rightarrow Na_2S(aq) + H_2O(l)$$

A student who was asked to balance this equation wrote the following:

$$Na_2OH(aq) + H_2S(g) \rightarrow Na_2S(aq) + H_3O(l)$$

Is this equation balanced? Is it correct? Explain why or why not, and supply the correct balanced equation if necessary.

It is balanced but incorrect. In two of the formulas the subscripts were changed, which changed the compounds involved. Water is not H_3O, and sodium hydroxide is not Na_2OH. The correct balanced equation is $2NaOH + H_2S \rightarrow Na_2S + 2H_2O$.

PROBLEM Write the answer on the line to the left. Show all your work in the space provided.

8. Recall that coefficients in a balanced chemical equation give relative amounts of moles as well as numbers of molecules.

_____30 mol_____ **a.** Calculate the number of moles of CO_2 that form if 10 mol of C_3H_4 react according to the following balanced equation:

$$C_3H_4 + 4O_2 \rightarrow 3CO_2 + 2H_2O$$

_____40 mol_____ **b.** Calculate the number of moles of O_2 that are consumed.

Name _____ Date _____ Class _____

CHAPTER 8 REVIEW
Chemical Equations and Reactions

SECTION 2

SHORT ANSWER Answer the following questions in the space provided.

1. Match the equation type on the left to its representation on the right.

   __c__ synthesis (a) $AX + BY \rightarrow AY + BX$

   __d__ decomposition (b) $A + BX \rightarrow AX + B$

   __b__ single-displacement (c) $A + B \rightarrow AX$

   __a__ double-displacement (d) $AX \rightarrow A + X$

2. __c__ In the reaction described by the equation $2Al(s) + 3Fe(NO_3)_2(aq) \rightarrow 3Fe(s) + 2Al(NO_3)_3(aq)$, iron has been replaced by

   (a) nitrate. (c) aluminum.
   (b) water. (d) nitrogen.

3. __a__ Of the following chemical equations, the only reaction that is both synthesis and combustion is

   (a) $C(s) + O_2(g) \rightarrow CO_2(g)$.
   (b) $2C_4H_{10}(l) + 13O_2(g) \rightarrow 8CO_2(g) + 10H_2O(l)$.
   (c) $6CO_2(g) + 6H_2O(g) \rightarrow C_6H_{12}O_6(aq) + 6O_2(g)$.
   (d) $C_6H_{12}O_6(aq) + 6O_2(g) \rightarrow 6CO_2(aq) + 6H_2O(l)$.

4. __b__ Of the following chemical equations, the only reaction that is both combustion and decomposition is

   (a) $S(s) + O_2(g) \rightarrow SO_2(g)$.
   (b) $2C_4H_{10}(l) + 13O_2(g) \rightarrow 8CO_2(g) + 10H_2O(l)$.
   (c) $2H_2O_2(l) \rightarrow 2H_2O(l) + O_2(g)$.
   (d) $2HgO(s) \xrightarrow{\Delta} 2Hg(l) + O_2(g)$.

5. Identify the products when the following substances decompose:

   __its separate elements__ a. a binary compound

   __metal oxide + water__ b. most metal hydroxides

   __metal oxide + carbon dioxide__ c. a metal carbonate

   __water + sulfur dioxide__ d. the acid H_2SO_3

6. The complete combustion of a hydrocarbon in excess oxygen yields the products __CO_2__ and __H_2O__.

Name _____ Date _____ Class _____

SECTION 2 continued

7. For the following four reactions, identify the type, predict the products (make sure formulas are correct), and balance the equations:

   a. $Cl_2(aq) + NaI(aq) \rightarrow$

   single-displacement; $Cl_2(aq) + 2NaI(aq) \rightarrow I_2(aq) + 2NaCl(aq)$

   b. $Mg(s) + N_2(g) \rightarrow$

   synthesis; $3Mg(s) + N_2(g) \rightarrow Mg_3N_2(s)$

   c. $Co(NO_3)_2(aq) + H_2S(aq) \rightarrow$

   double-displacement; $Co(NO_3)_2(aq) + H_2S(aq) \rightarrow CoS(s) + 2HNO_3(aq)$

   d. $C_2H_5OH(aq) + O_2(g) \rightarrow$

   combustion; $C_2H_5OH(aq) + 3O_2(g) \rightarrow 2CO_2(g) + 3H_2O(l)$

8. Acetylene gas, C_2H_2, is burned to provide the high temperature needed in welding.

   a. Write the balanced chemical equation for the combustion of C_2H_2 in oxygen.

   $2C_2H_2(g) + 5O_2(g) \rightarrow 4CO_2(g) + 2H_2O(l)$

   __2.0 mol__ b. If 1.0 mol of C_2H_2 is burned, how many moles of CO_2 are formed?

   __2.5 mol__ c. If 1.0 mol of C_2H_2 is burned how many moles of oxygen gas are consumed?

9. a. Write the balanced chemical equation for the reaction that occurs when solutions of barium chloride and sodium carbonate are mixed. Refer to **Table 1** on page 437 in **Chapter 13** for solubility.

   $BaCl_2(aq) + Na_2CO_3(aq) \rightarrow BaCO_3(s) + 2NaCl(aq)$

   b. To which of the five basic types of reactions does this reaction belong?

   double-displacement

10. For the commercial preparation of aluminum metal, the metal is extracted by electrolysis from alumina, Al_2O_3. Write the balanced chemical equation for the electrolysis of molten Al_2O_3.

    $2Al_2O_3(l) \rightarrow 4Al(s) + 3O_2(g)$

CHAPTER 8 REVIEW
Chemical Equations and Reactions

SECTION 3

SHORT ANSWER Answer the following questions in the space provided.

1. List four metals that will *not* replace hydrogen in an acid.
   Choose from Cu, Ag, Au, Pt, Sb, Bi, and Hg.

2. Consider the metals iron and silver, both listed in **Table 3** on page 286 of the text. Which one readily forms an oxide in nature, and which one does not?
   Fe forms an oxide in nature, and Ag does not, because it is much less active.

3. In each of the following pairs, identify the more active element.

   __F_2__ a. F_2 and I_2

   __K__ b. Mn and K

   __H__ c. Cu and H

4. Use the information in **Table 3** on page 286 of the text to predict whether each of the following reactions will occur. For each reaction that will occur, complete the chemical equation by writing in the products formed and balancing the final equation.

   a. $Al(s) + CH_3COOH(aq) \xrightarrow{50°C}$

   $2Al(s) + 6CH_3COOH(aq) \xrightarrow{50°C} 2Al(CH_3COO)_3(aq) + 3H_2(g)$

   b. $Al(s) + H_2O(l) \xrightarrow{50°C}$

   no reaction

   c. $Cr(s) + CdCl_2(aq) \rightarrow$

   $2Cr(s) + 3CdCl_2(aq) \rightarrow 2CrCl_3(aq) + 3Cd(s)$

   d. $Br_2(l) + KCl(aq) \rightarrow$

   no reaction

SECTION 3 continued

5. Very active metals will react with water to release hydrogen gas and form hydroxides.

   a. Complete, and then balance, the equation for the reaction of Ca(s) with water.

   $Ca(s) + 2H_2O(l) \rightarrow Ca(OH)_2(aq) + H_2(g)$

   b. The reaction of rubidium, Rb, with water is faster and more violent than the reaction of Na with water. Use the atomic structure and radius of each metal to account for this difference.

   Both are alkali metals and readily form a stable 1+ ion by ejecting an s^1 electron.

   Rb has a larger radius than Na and holds its electron less tightly, making it more reactive.

6. Gold, Au, is often used in jewelry. How does the relative activity of Au relate to its use in jewelry?
   Gold has a low reactivity and therefore does not corrode over time.

7. Explain how to use an activity series to predict the outcome of a single-displacement reaction.
   In single-displacement reactions, if the activity of the free element is greater than that of the element in the compound, the reaction will take place.

8. Aluminum is above copper in the activity series. Will aluminum metal react with copper(II) nitrate, $Cu(NO_3)_2$, to form aluminum nitrate, $Al(NO_3)_3$? If so, write the balanced chemical equation for the reaction.

   Yes; because aluminum is above copper in the activity series, aluminum metal will replace copper in copper(II) nitrate.

   $2Al(s) + 3Cu(NO_3)_2(aq) \rightarrow 2Al(NO_3)_3(aq) + 3Cu(s)$

Name _____ Date _____ Class _____

CHAPTER 8 REVIEW
Chemical Equations and Reactions

MIXED REVIEW

SHORT ANSWER Answer the following questions in the space provided.

1. __b__ A balanced chemical equation represents all the following *except*
    - (a) experimentally established facts.
    - (b) the mechanism by which reactants combine to form products.
    - (c) identities of reactants and products in a chemical reaction.
    - (d) relative quantities of reactants and products in a chemical reaction.

2. __d__ According to the law of conservation of mass, the total mass of the reacting substances is
    - (a) always more than the total mass of the products.
    - (b) always less than the total mass of the products.
    - (c) sometimes more and sometimes less than the total mass of the products.
    - (d) always equal to the total mass of the products.

3. Predict whether each of the following chemical reactions will occur. For each reaction that will occur, identify the reaction type and complete the chemical equation by writing in the products formed and balancing the final equation. General solubility rules are in **Table 1** on page 437 of the text.

    a. $Ba(NO_3)_2(aq) + Na_3PO_4(aq) \rightarrow$

    double-displacement; $3Ba(NO_3)_2(aq) + 2Na_3PO_4(aq) \rightarrow Ba_3(PO_4)_2(s) + 6NaNO_3(aq)$

    b. $Al(s) + O_2(g) \rightarrow$

    synthesis; $4Al(s) + 3O_2(g) \rightarrow 2Al_2O_3(s)$

    c. $I_2(s) + NaBr(aq) \rightarrow$

    no reaction

    d. $C_3H_4(g) + O_2(g) \rightarrow$

    combustion; $C_3H_4(g) + 4O_2(g) \rightarrow 3CO_2(g) + 2H_2O(g)$

MODERN CHEMISTRY   CHEMICAL EQUATIONS AND REACTIONS

MIXED REVIEW continued

    e. electrolysis of molten potassium chloride

    decomposition; $2KCl(l) \rightarrow 2K(s) + Cl_2(g)$

4. Some small rockets are powered by the reaction represented by the following unbalanced equation:

$$(CH_3)_2N_2H_2(l) + N_2O_4(g) \rightarrow N_2(g) + H_2O(g) + CO_2(g) + heat$$

    a. Translate this chemical equation into a sentence. (Hint: The name for $(CH_3)_2N_2H_2$ is dimethylhydrazine.)

    When liquid dimethylhydrazine is mixed with dinitrogen tetroxide gas, the products are nitrogen gas, water vapor, and gaseous carbon dioxide, along with energy in the form of heat.

    b. Balance the formula equation.

    $(CH_3)_2N_2H_2(l) + 2N_2O_4(g) \rightarrow 3N_2(g) + 4H_2O(g) + 2CO_2(g)$

5. In the laboratory, you are given two small chips of each of the unknown metals X, Y, and Z, along with dropper bottles containing solutions of $XCl_2(aq)$ and $ZCl_2(aq)$. Describe an experimental strategy you could use to determine the relative activities of X, Y, and Z.

    Wording and strategies will vary. First, place one chip of Y into $XCl_2(aq)$ and another into $ZCl_2(aq)$. If Y reacts with one solution but not the other, the activity series can be established. If Y replaces X but not Z, the series is $Z > Y > X$. If Y replaces Z but not X, the series is $X > Y > Z$. If Y reacts with neither solution, Y is at the bottom of the series. Next, put one chip of X into $ZCl_2(aq)$. If it reacts, the series is $X > Z > Y$. If it does not react, the series is $Z > X > Y$. If Y reacts with both solutions, Y is the most reactive. Last, put a chip of X into $ZCl_2(aq)$. If it reacts, the series is $Y > X > Z$. If it does not react, the series is $Y > Z > X$.

6. List the observations that would indicate that a reaction had occurred.

    Signs of a reaction include generation of energy as heat or light, formation of a precipitate, formation of a gas, and change in color.

Name _____ Date _____ Class _____

CHAPTER 9 REVIEW
Stoichiometry

SECTION 1

SHORT ANSWER Answer the following questions in the space provided.

1. __b__ The coefficients in a chemical equation represent the
   - (a) masses in grams of all reactants and products.
   - (b) relative number of moles of reactants and products.
   - (c) number of atoms of each element in each compound in a reaction.
   - (d) number of valence electrons involved in a reaction.

2. __d__ Which of the following would not be studied within the topic of stoichiometry?
   - (a) the mole ratio of Al to Cl in the compound aluminum chloride
   - (b) the mass of carbon produced when a known mass of sucrose decomposes
   - (c) the number of moles of hydrogen that will react with a known quantity of oxygen
   - (d) the amount of energy required to break the ionic bonds in CaF_2

3. __a__ A balanced chemical equation allows you to determine the
   - (a) mole ratio of any two substances in the reaction.
   - (b) energy released in the reaction.
   - (c) electron configuration of all elements in the reaction.
   - (d) reaction mechanism involved in the reaction.

4. __c__ The relative number of moles of hydrogen to moles of oxygen that react to form water represents a(n)
   - (a) reaction sequence.
   - (b) bond energy.
   - (c) mole ratio.
   - (d) element proportion.

5. Given the reaction represented by the following unbalanced equation: $N_2O(g) + O_2(g) \rightarrow NO_2(g)$

   a. Balance the equation.

   __$2N_2O(g) + 3O_2(g) \rightarrow 4NO_2(g)$__

   __4 mol NO_2:3 mol O_2__ b. What is the mole ratio of NO_2 to O_2?

   __15.0 mol__ c. If 20.0 mol of NO_2 form, how many moles of O_2 must have been consumed?

   __True__ d. Twice as many moles of NO_2 form as moles of N_2O are consumed. True or False?

   __False__ e. Twice as many grams of NO_2 form as grams of N_2O are consumed. True or False?

MODERN CHEMISTRY STOICHIOMETRY **73**

SECTION 1 continued

PROBLEMS Write the answer on the line to the left. Show all your work in the space provided.

6. Given the following equation: $N_2(g) + 3H_2(g) \rightarrow 2NH_3(g)$

__28.0 g/mol N_2__   **a.** Determine to one decimal place the molar mass of each substance and express each mass in grams per mole.

__2.0 g/mol H_2__

__17.0 g/mol NH_3__

b. There are six different mole ratios in this system. Write out each one.

__3 mol H_2:1 mol N_2; 2 mol NH_3:1 mol N_2; 2 mol NH_3:3 mol H_2; or their reciprocals__

7. Given the following equation: $4NH_3(g) + 6NO(g) \rightarrow 5N_2(g) + 6H_2O(g)$

__1 mol NO:1 mol H_2O__   **a.** What is the mole ratio of NO to H_2O?

__3 mol NO:2 mol NH_3__   **b.** What is the mole ratio of NO to NH_3?

__0.360 mol__   **c.** If 0.240 mol of NH_3 react according to the above equation, how many moles of NO will be consumed?

8. Propyne gas can be used as a fuel. The combustion reaction of propyne can be represented by the following equation:

$$C_3H_4(g) + 4O_2(g) \rightarrow 3CO_2(g) + 2H_2O(g)$$

a. Write all the possible mole ratios in this system.

__4 mol O_2:1 mol C_3H_4; 3 mol CO_2:1 mol C_3H_4; 2 mol H_2O:1 mol C_3H_4;__

__3 mol CO_2:4 mol O_2; 2 mol H_2O:4 mol O_2; 2 mol H_2O:3 mol CO_2;__

__or their reciprocals__

b. Suppose that x moles of water form in the above reaction. The other three mole quantities (*not* in order) are $2x$, $1.5x$, and $0.5x$. Match these quantities to their respective components in the equation above.

__C_3H_4 is $0.5x$; O_2 is $2x$; and CO_2 is $1.5x$__

CHAPTER 9 REVIEW
Stoichiometry

SECTION 2

PROBLEMS Write the answer on the line to the left. Show all your work in the space provided.

1. _____4.5 mol_____ The following equation represents a laboratory preparation for oxygen gas:

    $2KClO_3(s) \rightarrow 2KCl(s) + 3O_2(g)$

    How many moles of O_2 form if 3.0 mol of $KClO_3$ are totally consumed?

2. _____200 g_____ Given the following equation: $H_2(g) + F_2(g) \rightarrow 2HF(g)$
    How many grams of HF gas are produced as 5 mol of fluorine react?

3. _____0.53 g_____ Water can be made to decompose into its elements by using electricity according to the following equation:

    $2H_2O(l) \rightarrow 2H_2(g) + O_2(g)$

    How many grams of O_2 are produced when 0.033 mol of water decompose?

4. _____34.8 g_____ Sodium metal reacts with water to produce NaOH according to the following equation:

    $2Na(s) + 2H_2O(l) \rightarrow 2NaOH(aq) + H_2(g)$

    How many grams of NaOH are produced if 20.0 g of sodium metal react with excess oxygen?

Name _____ Date _____ Class _____

SECTION 2 continued

5. _____ 60.2 g _____ a. What mass of oxygen gas is produced if 100. g of lithium perchlorate are heated and allowed to decompose according to the following equation?

$$LiClO_4(s) \rightarrow LiCl(s) + 2O_2(g)$$

_____ 42.1 L _____ b. The oxygen gas produced in part **a** has a density of 1.43 g/L. Calculate the volume of this gas.

6. A car air bag requires 70. L of nitrogen gas to inflate properly. The following equation represents the production of nitrogen gas:

$$2NaN_3(s) \rightarrow 2Na(s) + 3N_2(g)$$

_____ 81 g _____ a. The density of nitrogen gas is typically 1.16 g/L at room temperature. Calculate the number of grams of N_2 that are needed to inflate the air bag.

_____ 2.9 mol _____ b. Calculate the number of moles of N_2 that are needed.

_____ 1.3×10^2 g _____ c. Calculate the number of grams of NaN_3 that must be used to generate the amount of N_2 necessary to properly inflate the air bag.

Name _____ Date _____ Class _____

CHAPTER 9 REVIEW
Stoichiometry

SECTION 3

PROBLEMS Write the answer on the line to the left. Show all your work in the space provided.

1. _____88%_____ The actual yield of a reaction is 22 g and the theoretical yield is 25 g. Calculate the percentage yield.

2. 6.0 mol of N_2 are mixed with 12.0 mol of H_2 according to the following equation:

$$N_2(g) + 3H_2(g) \rightarrow 2NH_3(g)$$

_____N_2; 2.0 mol_____ a. Which chemical is in excess? What is the excess in moles?

_____8.0 mol_____ b. Theoretically, how many moles of NH_3 will be produced?

_____6.4 mol_____ c. If the percentage yield of NH_3 is 80%, how many moles of NH_3 are actually produced?

3. 0.050 mol of $Ca(OH)_2$ are combined with 0.080 mol of HCl according to the following equation:

$$Ca(OH)_2(aq) + 2HCl(aq) \rightarrow CaCl_2(aq) + 2H_2O(l)$$

_____0.10 mol_____ a. How many moles of HCl are required to neutralize all 0.050 mol of $Ca(OH)_2$?

MODERN CHEMISTRY   STOICHIOMETRY

SECTION 3 continued

_____HCl_____ b. What is the limiting reactant in this neutralization reaction?

_____1.4 g_____ c. How many grams of water will form in this reaction?

4. Acid rain can form in a two-step process, producing $HNO_3(aq)$.

$$N_2(g) + 2O_2(g) \rightarrow 2NO_2(g)$$
$$3NO_2(g) + H_2O(g) \rightarrow 2HNO_3(aq) + NO(g)$$

_____1.26×10^3 g_____ a. A car burns 420. g of N_2 according to the above equations. How many grams of HNO_3 will be produced?

_____960. g_____ b. For the above reactions to occur, O_2 must be in excess in the first step. What is the minimum amount of O_2 needed in grams?

_____6.9×10^2 L_____ c. What volume does the amount of O_2 in part b occupy if its density is 1.4 g/L?

CHAPTER 9 REVIEW
Stoichiometry

MIXED REVIEW

SHORT ANSWER Answer the following questions in the space provided.

1. Given the following equation: $C_3H_4(g) + xO_2(g) \rightarrow 3CO_2(g) + 2H_2O(g)$

    __4__ a. What is the value of the coefficient x in this equation?

    __40.07 g/mol__ b. What is the molar mass of C_3H_4?

    __2 mol O_2 : 1 mol H_2O__ c. What is the mole ratio of O_2 to H_2O in the above equation?

    __0.20 mol__ d. How many moles are in an 8.0 g sample of C_3H_4?

    __3z__ e. If z mol of C_3H_4 react, how many moles of CO_2 are produced, in terms of z?

2. a. What is meant by *ideal conditions* relative to stoichiometric calculations?

    The limiting reactant is completely converted to product with no losses, as dictated by the ratio of coefficients.

    b. What function do ideal stoichiometric calculations serve?

    They determine the theoretical yield of the products of the reaction.

    c. Are actual yields typically larger or smaller than theoretical yields?

    smaller

PROBLEMS Write the answer on the line to the left. Show all your work in the space provided.

3. Assume the reaction represented by the following equation goes all the way to completion:

    $$N_2 + 3H_2 \rightarrow 2NH_3$$

    __4 mol__ a. If 6 mol of H_2 are consumed, how many moles of NH_3 are produced?

    __8.5 g__ b. How many grams are in a sample of NH_3 that contains 3.0×10^{23} molecules?

MIXED REVIEW continued

c. If 0.1 mol of N_2 combine with H_2, what must be true about the quantity of H_2 for N_2 to be the limiting reactant?

At least 0.3 mol of H_2 must be provided.

4. __75%__ If a reaction's theoretical yield is 8.0 g and the actual yield is 6.0 g, what is the percentage yield?

5. Joseph Priestley generated oxygen gas by strongly heating mercury(II) oxide according to the following equation:

$$2HgO(s) \rightarrow 2Hg(l) + O_2(g)$$

__0.0693 mol__ a. If 15.0 g HgO decompose, how many moles of HgO does this represent?

__0.0346 mol__ b. How many moles of O_2 are theoretically produced?

__1.11 g__ c. How many grams of O_2 is this?

__0.786 L__ d. If the density of O_2 gas is 1.41 g/L, how many liters of O_2 are produced?

__1.05 g__ e. If the percentage yield is 95.0%, how many grams of O_2 are actually collected?

CHAPTER 10 REVIEW

States of Matter

SECTION 1

SHORT ANSWER Answer the following questions in the space provided.

1. Identify whether the descriptions below describe an ideal gas or a real gas.

   __ideal gas__ a. The gas will not condense because the molecules do not attract each other.

   __ideal gas__ b. Collisions between molecules are perfectly elastic.

   __real gas__ c. Gas particles passing close to one another exert an attraction on each other.

2. The formula for kinetic energy is $KE = \frac{1}{2}mv^2$.

   a. As long as temperature is constant, what happens to the kinetic energy of the colliding particles during an elastic collision?

   __The energy is transferred between them.__

   b. If two gases have the same temperature and share the same energy but have different molecular masses, which molecules will have the greater speed?
   __Those with the lower molecule mass.__

3. Use the kinetic-molecular theory to explain each of the following phenomena:

   a. A strong-smelling gas released from a container in the middle of a room is soon detected in all areas of that room.

   __Gas molecules are in constant, rapid, random motion.__

   b. As a gas is heated, its rate of effusion through a small hole increases if all other factors remain constant.

   __As a gas is heated, each molecule's speed increases; therefore, the molecules pass through the small hole more frequently.__

4. a. __b, d, c, a__ List the following gases in order of rate of effusion, from lowest to highest. (Assume all gases are at the same temperature and pressure.)

   (a) He   (b) Xe   (c) HCl   (d) Cl_2

SECTION 1 continued

b. Explain why you put the gases in the order above. Refer to the kinetic-molecular theory to support your explanation.

All gases at the same temperature have the same average kinetic energy. Therefore, heavier molecules have slower average speeds. Thus, the gases are ranked from heaviest to lightest in molar mass.

5. Explain why polar gas molecules experience larger deviations from ideal behavior than nonpolar molecules when all other factors (mass, temperature, etc) are held constant.

Polar molecules attract neighboring polar molecules and often move out of their straight-line paths because of these attractions.

6. __c__ The two gases in the figure below are simultaneously injected into opposite ends of the tube. The ends are then sealed. They should just begin to mix closest to which labeled point?

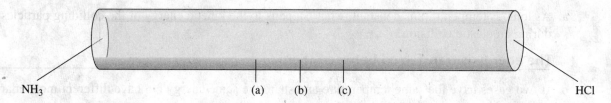

7. Explain the difference in the speed-distribution curves of a gas at the two temperatures shown in the figure below.

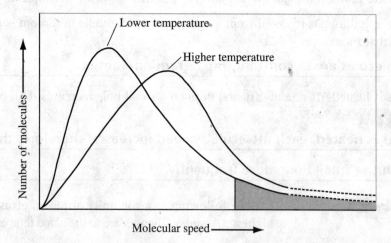

In both cases the average speed of the molecules is proportional to temperature. The distribution of molecules becomes broader as the temperature increases. This means that there are a greater number of molecules traveling within a greater range of higher speeds as the temperature increases.

Name _____ Date _____ Class _____

CHAPTER 10 REVIEW

States of Matter

SECTION 2

SHORT ANSWER Answer the following questions in the space provided.

1. __a__ Liquids possess all the following properties *except*

    (a) relatively low density.
    (b) the ability to diffuse.
    (c) relative incompressibility.
    (d) the ability to change to a gas.

2. a. Chemists distinguish between intermolecular and intramolecular forces. Explain the difference between these two types of forces.

   Intermolecular forces are between separate molecules; intramolecular forces are

   within individual molecules.

   Classify each of the following as intramolecular or intermolecular:

   __intermolecular__ b. hydrogen bonding in liquid water

   __intramolecular__ c. the O—H covalent bond in methanol, CH_3OH

   __intermolecular__ d. the bonds that cause gaseous Cl_2 to become a liquid when cooled

3. Explain the following properties of liquids by describing what is occurring at the molecular level.

   a. A liquid takes the shape of its container but does not expand to fill its volume.

   Liquid molecules are very mobile. This mobility allows a liquid to take the shape of

   its container. In liquids, molecules are in contact with adjacent molecules, allowing

   intermolecular forces to have a greater effect than they do in gases. The molecules in

   a liquid will therefore not necessarily spread out to fill a container's entire volume.

   b. Polar liquids are slower to evaporate than nonpolar liquids.

   Polar molecules are attracted to adjacent molecules and are therefore less able to

   escape from the liquid's surface than are nonpolar molecules.

MODERN CHEMISTRY

SECTION 2 continued

4. Explain briefly why liquids tend to form spherical droplets, decreasing surface area to the smallest size possible.

   An attractive force pulls adjacent parts of a liquid's surface together, thus decreasing surface area to the smallest possible size. A sphere offers the minimum surface area for a given volume of liquid.

5. Is freezing a chemical change or a physical change? Briefly explain your answer.

   Freezing is a physical change. The substance solidifying is changing its state, which is a physical change. It is still the same substance so it has not changed chemically.

6. Is evaporation a chemical or physical change? Briefly explain your answer.

   Evaporation is a physical change because it involves a change of physical state. There is no change in the chemical makeup of the substance, which would be necessary for a chemical change.

7. What is the relationship between vaporization and evaporation?

   Evaporation is a form of vaporization. It occurs only in nonboiling liquids when some liquid particles enter the gas state. Vaporization is a more general term that refers to either a liquid or a solid changing to a gas.

CHAPTER 10 REVIEW
States of Matter

SECTION 3

SHORT ANSWER Answer the following questions in the space provided.

1. Match description on the right to the correct crystal type on the left.

   __b__ ionic crystal (a) has mobile electrons in the crystal

   __c__ covalent molecular crystal (b) is hard, brittle, and nonconducting

   __a__ metallic crystal (c) typically has the lowest melting point of the four crystal types

   __d__ covalent network crystal (d) has strong covalent bonds between neighboring atoms

2. For each of the four types of solids, give a specific example other than one listed in Table 1 on page 340 of the text.

   some possible answers:

   ionic solid: MgO, CaO, KI, $CuSO_4$

   covalent network solid: graphite, silicon carbide

   covalent molecular solid: dry ice (CO_2), sulfur, iodine

   metallic solid: any metal from the far left side of the periodic table

3. A chunk of solid lead is dropped into a pool of molten lead. The chunk sinks to the bottom of the pool. What does this tell you about the density of the solid lead compared with the density of the molten lead?

   Solid lead is denser than the liquid form.

4. Answer *amorphous solid* or *crystalline solid* to the following questions:

   __crystalline solid__ a. Which is less compressible?

   __crystalline solid__ b. Which has a more clearly defined shape?

   __amorphous solid__ c. Which is sometimes described as a supercooled liquid?

   __amorphous solid__ d. Which has a less clearly defined melting point?

SECTION 3 continued

5. Explain the following properties of solids by describing what is occurring at the atomic level.

   a. Metallic solids conduct electricity well, but covalent network solids do not.

   Metals have many electrons that are not bound to any one atom; therefore they are able to move throughout the crystal. In covalent network solids, all atoms (and electrons) are strongly bound in place and are not free to move.

   b. The volume of a solid changes only slightly with a change in temperature or pressure.

   Solids have definite volume because their particles are packed very close together. There is very little empty space into which the particles can be compressed. Even at high temperatures their particles are held in relatively fixed positions.

   c. Amorphous solids do not have a definite melting point.

   In amorphous solids, particles are arranged randomly; no specific amount of kinetic energy is needed to overcome the attractive forces holding the particles together. Thus, they do not have a point at which they melt, but melt over a range of temperatures.

   d. Ionic crystals are much more brittle than covalent molecular crystals.

   Ionic crystals have strong binding forces between the positive and negative ions in the crystal structure. Covalent molecular crystals have weaker bonds between the molecules.

6. Experiments show that it takes 6.0 kJ of energy to melt 1 mol of water ice at its melting point but only about 1.1 kJ to melt 1 mol of methane, CH_4, at its melting point. Explain in terms of intermolecular forces why it takes so much less energy to melt the methane.

   The attractive forces between CH_4 molecules are weak (dispersion forces). Little energy is needed to separate the molecules. Melting water ice involves the breaking of many hydrogen bonds between molecules, which requires more energy.

CHAPTER 10 REVIEW
States of Matter

SECTION 4

SHORT ANSWER Answer the following questions in the space provided.

1. __a__ When a substance in a closed system undergoes a phase change and the system reaches equilibrium,
    - (a) the two opposing changes occur at equal rates.
    - (b) there are no more phase changes.
    - (c) one phase change predominates.
    - (d) the amount of substance in the two phases changes.

2. Match the following definitions on the right with the words on the left.

    __b__ equilibrium   (a) melting

    __c__ volatile   (b) opposing changes occurring at equal rates in a closed system

    __a__ fusion   (c) readily evaporated

    __d__ deposition   (d) a change directly from a gas to a solid

3. Match the process on the right with the change of state on the left.

    __c__ solid to gas   (a) melting

    __d__ liquid to gas   (b) condensation

    __b__ gas to liquid   (c) sublimation

    __a__ solid to liquid   (d) vaporization

4. Refer to the phase diagram for water in Figure 16 on page 347 of the text to answer the following questions:

    __A__ a. What point represents the conditions under which all three phases can coexist?

    __C__ b. What point represents a temperature above which only the vapor phase exists?

    __it decreases__ c. Based on the diagram, as the pressure on the water system increases, what happens to the melting point of ice?

    d. What happens when water is at point A on the curve and the temperature increases while the pressure is held constant?
    ice and liquid water will vaporize, forming water vapor

SECTION 4 continued

5. Use this general equilibrium equation to answer the following questions:

$$\text{reactants} \rightleftarrows \text{products} + \text{energy}$$

__decrease__ **a.** If the forward reaction is favored, will the concentration of reactants increase, decrease, or stay the same?

__reverse reaction__ **b.** If extra product is introduced, which reaction will be favored?

__forward reaction__ **c.** If the temperature of the system decreases, which reaction will be favored?

6. Refer to the graph below to answer the following questions:

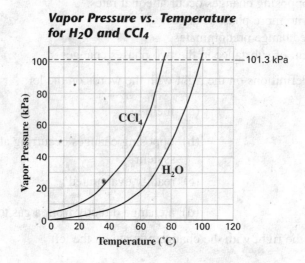

Vapor Pressure vs. Temperature for H₂O and CCl₄

__about 75°C__ **a.** What is the normal boiling point of CCl_4?

__about 85°C__ **b.** What would be the boiling point of water if the air pressure over the liquid were reduced to 60 kPa?

__about 38 kPa__ **c.** What must the air pressure over CCl_4 be for it to boil at 50°C?

d. Although water has a lower molar mass than CCl_4, it has a lower vapor pressure when measured at the same temperature. What makes water vapor less volatile than CCl_4?

Based solely on molar mass, CCl₄ would be expected to be less volatile than water.

However, CCl₄ is nonpolar and thus has weak intermolecular forces of attraction.

Water is polar and contains strong hydrogen bonds between molecules. Thus,

water is less volatile despite its smaller molar mass.

CHAPTER 10 REVIEW
States of Matter

SECTION 5

SHORT ANSWER Answer the following questions in the space provided.

1. Indicate whether each of the following is a *physical* or *chemical* property of water.

    __physical__ a. The density of ice is less than the density of liquid water.

    __chemical__ b. A water molecule contains one atom of oxygen and two atoms of hydrogen.

    __chemical__ c. There are strong hydrogen bonds between water molecules.

    __physical__ d. Ice consists of water molecules in a hexagonal arrangement.

2. Compare a polar water molecule with a less-polar molecule, such as formaldehyde, CH_2O. Both are liquids at room temperature and 1 atm pressure.

    __water__ a. Which liquid should have the higher boiling point?

    __formaldehyde__ b. Which liquid is more volatile?

    __water__ c. Which liquid has a higher surface tension?

    __water__ d. In which liquid is NaCl, an ionic crystal, likely to be more soluble?

3. Describe hydrogen bonding as it occurs in water in terms of the location of the bond, the particles involved, the strength of the bond, and the effects this type of bonding has on physical properties.

    Hydrogen bonding in water occurs between a hydrogen atom of one water molecule and the unshared pair of electrons of an oxygen atom of an adjacent water molecule. It is a particularly strong type of dipole-dipole force. Hydrogen bonding causes the boiling point of water and its molar enthalpy of vaporization to be relatively high. The water's high surface tension is also a result of hydrogen bonding.

SECTION 5 continued

PROBLEMS Write the answer on the line to the left. Show all your work in the space provided.

4. The molar enthalpy of vaporization of water is 40.79 kJ/mol, and the molar enthalpy of fusion of ice is 6.009 kJ/mol. The molar mass of water is 18.02 g/mol.

___68.6 kJ/mol___ **a.** How much energy is absorbed when 30.3 g of liquid water boils?

___79.8 cal/g___ **b.** An energy unit often encountered is the calorie (4.18 J = 1 calorie). Determine the molar enthalpy of fusion of ice in calories per gram.

5. A typical ice cube has a volume of about 16.0 cm^3. Calculate the amount of energy needed to melt the ice cube. (Density of ice at 0.°C = 0.917 g/mL; molar enthalpy of fusion of ice = 6.009 kJ/mol; molar mass of H$_2$O = 18.02 g/mol.)

___14.7 g___ **a.** Determine the mass of the ice cube.

___0.814 mol___ **b.** Determine the number of moles of H$_2$O present in the sample.

___4.89 kJ___ **c.** Determine the number of kilojoules of energy needed to melt the ice cube.

CHAPTER 10 REVIEW

States of Matter

MIXED REVIEW

SHORT ANSWER Answer the following questions in the space provided.

1. __c__ The average speed of a gas molecule is most directly related to the
   - (a) polarity of the molecule.
   - (b) pressure of the gas.
   - (c) temperature of the gas.
   - (d) number of moles in the sample.

2. Use the kinetic-molecular theory to explain the following phenomena:

   a. When 1 mol of a real gas is condensed to a liquid, the volume shrinks by a factor of about 1000.

   Molecules in a gas are far apart. They are much closer together in a liquid.

   Molecules in a gas are easily squeezed closer together as the gas is compressed.

   b. When a gas in a rigid container is warmed, the pressure on the walls of the container increases.

   As the temperature increases, the molecules speed up. Thus, they collide with the walls more frequently than before and with a greater force per impact. For both of these reasons, the total force per unit area increases and the pressure increases.

3. __b__ Which of the following statements about liquids and gases is *not* true?
   - (a) Molecules in a liquid are much more closely packed than molecules in a gas.
   - (b) Molecules in a liquid can vibrate and rotate, but they are bound in fixed positions.
   - (c) Liquids are much more difficult to compress into a smaller volume than are gases.
   - (d) Liquids diffuse more slowly than gases.

4. Answer *solid* or *liquid* to the following questions:

   __solid__ a. Which is less compressible?

   __liquid__ b. Which is quicker to diffuse into neighboring media?

   __solid__ c. Which has a definite volume and shape?

   __solid__ d. Which has molecules that are rotating or vibrating primarily in place?

MIXED REVIEW continued

5. Explain why almost all solids are denser than their liquid states by describing what is occurring at the molecular level.

In solids, particles are more closely packed than in liquids, due to stronger attractive

forces between the particles of the solid.

6. A general equilibrium equation for boiling is

$$\text{liquid} + \text{energy} \rightleftarrows \text{vapor}$$

Indicate whether the forward or reverse reaction is favored in each of the following cases:

__forward reaction__ a. The temperature of the system is increased.

__reverse reaction__ b. More molecules of the vapor are added to the system.

__reverse reaction__ c. The pressure on the system is increased.

7. __181 kJ__ Freon-11, CCl_3F has been commonly used in air conditioners. It has a molar mass of 137.35 g/mol and its enthalpy of vaporization is 24.8 kJ/mol at its normal boiling point of 24°C. Ideally how much energy in the form of heat is removed from a room by an air conditioner that evaporates 1.00 kg of freon-11?

8. Use the data table below to answer the following:

Composition	Molar mass (g/mol)	Enthalpy vaporization (kJ/mol)	Normal boiling point (°C)	Critical temperature (°C)
He	4	0.08	−269	−268
Ne	20	1.8	−246	−229
Ar	40	6.5	−186	−122
Xe	131	12.6	−107	+17
H_2O	18	40.8	+100	+374
HF	20	25.2	+20	+188
CH_4	16	8.9	−161	−82
C_2H_6	30	15.7	−89	+32

__higher__ a. Among *nonpolar* liquids, those with higher molar masses tend to have normal boiling points that are (higher, lower, or about the same).

__higher__ b. Among compounds of approximately the same molar mass, those with greater polarities tend to have enthalpies of vaporization that are (higher, lower, or about the same).

c. Which is the only noble gas listed that is stable as a liquid at 0°C? Explain your answer using the concept of critical temperature.

Xe; a substance can exist only as a gas at temperatures above its critical temperature.

Of the noble gases listed, only Xe has a critical temperature above 0°C.

Name _____ Date _____ Class _____

CHAPTER 11 REVIEW
Gases

SECTION 1

SHORT ANSWER Answer the following questions in the space provided.

1. __b__ $Pressure = \frac{force}{surface\ area}$. For a constant force, when the surface area is tripled the pressure is
   - (a) doubled.
   - (b) a third as much.
   - (c) tripled.
   - (d) unchanged.

2. __d, c, a, b__ Rank the following pressures in increasing order.
   - (a) 50 kPa
   - (b) 2 atm
   - (c) 76 torr
   - (d) 100 N/m²

3. Explain how to calculate the partial pressure of a dry gas that is collected over water when the total pressure is atmospheric pressure.

   Subtract the vapor pressure of water at the given collecting temperature from the

   atmospheric pressure taken during the collection of the gas.

PROBLEMS Write the answer on the line to the left. Show all your work in the space provided.

4. a. Use five to six data points from **Appendix Table A-8** in the text to sketch the curve for water vapor's partial pressure versus temperature on the graph provided below.

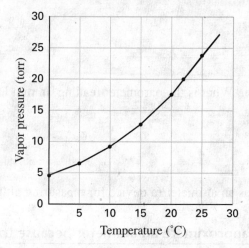

__No__ b. Do the data points lie on a straight line?

__10 torr__ c. Based on your sketch, predict the approximate partial pressure for water at 11°C.

SECTION 1 continued

5. Convert a pressure of 0.200 atm to the following units:

   _____152_____ a. mm Hg

   _____20.3_____ b. kPa

6. When an explosive like TNT is detonated, a mixture of gases at high temperature is created. Suppose that gas X has a pressure of 50 atm, gas Y has a pressure of 20 atm, and gas Z has a pressure of 10 atm.

   _____80 atm_____ a. What is the total pressure in this system?

   _____8.1×10^3 kPa_____ b. What is the total pressure in this system in kPa?

7. The height of the mercury in a barometer is directly proportional to the pressure on the mercury's surface. At sea level, pressure averages 1.0 atm and the level of mercury in the barometer is 760 mm (30. in.). In a hurricane, the barometric reading may fall to as low as 28 in.

   _____0.93 atm_____ a. Convert a pressure reading of 28 in. to atmospheres.

   _____3.8×10^2 mm Hg_____ b. What is the barometer reading, in mm Hg, at a pressure of 0.50 atm?

   c. Can a barometer be used as an altimeter (a device for measuring altitude above sea level)? Explain your answer.

   Yes; a barometer can approximate an altimeter because the higher you climb into

   Earth's atmosphere, the lower the pressure recorded by the barometer.

Name _____ Date _____ Class _____

CHAPTER 11 REVIEW

Gases

SECTION 2

SHORT ANSWER   Answer the following questions in the space provided.

1. State whether the pressure of a fixed mass of gas will increase, decrease, or stay the same in the following circumstances:

   __increase__ a. temperature increases, volume stays the same

   __decrease__ b. volume increases, temperature stays the same

   __decrease__ c. temperature decreases, volume stays the same

   __increase__ d. volume decreases, temperature stays the same

2. Two sealed flasks, A and B, contain two different gases of equal volume at the same temperature and pressure. Assume that flask A is warmed as flask B is cooled. Will the pressure in the two flasks remain equal? If not, which flask will have the higher pressure?

   No, the pressure will not remain equal. If all other factors remain constant, flask A

   will have the higher pressure because it is at the higher temperature.

PROBLEMS   Write the answer on the line to the left. Show all your work in the space provided.

3. A bicycle tire is inflated to 55 lb/in.^{2} at 15°C. Assume that the volume of the tire does not change appreciably once it is inflated.

   a. If the tire and the air inside it are heated to 30°C by road friction, does the pressure in the tire increase or decrease? (Assume the volume of air in the tire remains constant.)

   The pressure increases as the temperature rises (at fixed mass and constant

   volume).

   b. Because the temperature has doubled, does the pressure double to 110 psi?

   The pressure does not double.

   c. What will the pressure be when the temperature has doubled? Express your answer in pounds per square inch.

   57.9 psi

MODERN CHEMISTRY   GASES **95**

Name _____ Date _____ Class _____

SECTION 2 continued

4. _____4.5 atm_____ A 24 L sample of a gas at fixed mass and constant temperature exerts a pressure of 3.0 atm. What pressure will the gas exert if the volume is changed to 16 L?

5. _____16.7 mL_____ A common laboratory system to study Boyle's law uses a gas trapped in a syringe. The pressure in the system is changed by adding or removing identical weights on the plunger. The original gas volume is 50.0 mL when two weights are present. Predict the new gas volume when four more weights are added.

6. _____1.03 L_____ A sample of argon gas occupies a volume of 950 mL at 25.0°C. What volume will the gas occupy at 50.0°C if the pressure remains constant?

7. _____1.90×10^5 Pa_____ A 500.0 mL gas sample at STP is compressed to a volume of 300.0 mL and the temperature is increased to 35.0°C. What is the new pressure of the gas in pascals?

8. _____1.27×10^3 mL_____ A sample of gas occupies 1000. mL at standard pressure. What volume will the gas occupy at a pressure of 600. mm Hg if the temperature remains constant?

CHAPTER 11 REVIEW

Gases

SECTION 3

SHORT ANSWER Answer the following questions in the space provided.

1. __c__ The molar mass of a gas at STP is the density of that gas
   - (a) multiplied by the mass of 1 mol.
   - (b) divided by the mass of 1 mol.
   - (c) multiplied by 22.4 L.
   - (d) divided by 22.4 L.

2. __c__ For the expression $V = \frac{nRT}{P}$, which of the following will cause the volume to increase?
   - (a) increasing P
   - (b) decreasing T
   - (c) increasing T
   - (d) decreasing n

3. Two sealed flasks, A and B, contain two different gases of equal volume at the same temperature and pressure.

   __True__ a. The two flasks must contain an equal number of molecules. True or False?

   __False__ b. The two samples must have equal masses. True or False?

PROBLEMS Write the answer on the line to the left. Show all your work in the space provided.

4. Use the data in the table below to answer the following questions.

Formula	Molar mass (g/mol)
N_2	28.02
CO	28.01
C_2H_2	26.04
He	4.00
Ar	39.95

(Assume all gases are at STP.)

__all five gases__ a. Which gas contains the most molecules in a 5.0 L sample?

__He__ b. Which gas is the least dense?

__CO and N_2__ c. Which two gases have virtually the same density?

__1.25 g/L__ d. What is the density of N_2 measured at STP?

SECTION 3 continued

5. __0.25 mol__ a. How many moles of methane, CH_4 are present in 5.6 L of the gas at STP?

   __0.25 mol__ b. How many moles of gas are present in 5.6 L of any ideal gas at STP?

   __4.0 g__ c. What is the mass of the 5.6 L sample of CH_4?

6. __5.8 mol__ a. A large cylinder of He gas, such as that used to inflate balloons, has a volume of 25.0 L at 22°C and 5.6 atm. How many moles of He are in such a cylinder?

   __23 g__ b. What is the mass of the He calculated in part **a**?

7. When C_3H_4 combusts at STP, 5.6 L of C_3H_4 are consumed according to the following equation:

   $$C_3H_4(g) + 4O_2(g) \rightarrow 3CO_2(g) + 2H_2O(l)$$

   __0.25 mol__ a. How many moles of C_3H_4 react?

   __1.0 mol of O_2__
   __0.75 mol of CO_2__   b. How many moles of O_2, CO_2, and H_2O are either consumed or produced in the above reaction?
   __0.50 mol of H_2O__

   __10. g__ c. How many grams of C_3H_4 are consumed?

   __17 L__ d. How many liters of CO_2 are produced?

   __9.0 g__ e. How many grams of H_2O are produced?

CHAPTER 11 REVIEW

Gases

SECTION 4

SHORT ANSWER Answer the following questions in the space provided.

1. __b, d, c, a__ List the following gases in order of increasing rate of effusion. (Assume all gases are at the same temperature and pressure.)

   (a) He   (b) Xe   (c) HCl   (d) Cl_2

2. Explain your reasoning for the order of gases you chose in item **1** above. Refer to the kinetic-molecular theory to support your explanation and cite Graham's law of effusion.

   All gases at the same temperature have the same average kinetic energy. Therefore, heavier molecules have slower average speeds. Graham's law states that molecular speeds vary inversely with the square roots of their molar masses. Thus, the gases are ranked from heaviest to lightest in molar mass.

3. __c__ The two gases in the figure below are simultaneously injected into opposite ends of the tube. At which labeled point should they just begin to mix?

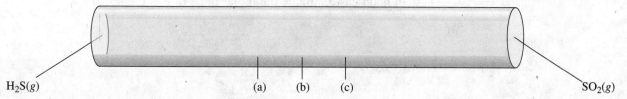

$H_2S(g)$ (a) (b) (c) $SO_2(g)$

4. State whether each example describes effusion or diffusion.

   __effusion__ a. As a puncture occurs, air moves out of a bicycle tire.

   __diffusion__ b. When ammonia is spilled on the floor, the house begins to smell like ammonia.

   __diffusion__ c. The smell of car exhaust pervades an emissions testing station.

5. Describe what happens, in terms of diffusion, when a bottle of perfume is opened.

   The molecules of the perfume randomly diffuse into the air and mix with the air molecules. At the same time, some molecules from the air diffuse into the bottle and mix with the perfume molecules.

Name _____ Date _____ Class _____

SECTION 4 continued

PROBLEMS Write the answer on the line to the left. Show all your work in the space provided.

6. _____1:9_____ a. The molar masses of He and of HCl are 4.00 g/mol and 36.46 g/mol, respectively. What is the ratio of the mass of He to the mass of HCl rounded to one decimal place?

_____3:1_____ b. Use your answer in part **a** to calculate the ratio of the average speed of He to the average speed of HCl.

_____400 m/s_____ c. If helium's average speed is 1200 m/s, what is the average speed of HCl?

7. _____202 g/mol_____ An unknown gas effuses through an opening at a rate 3.16 times slower than neon gas. Estimate the molar mass of this unknown gas.

CHAPTER 11 REVIEW

Gases

MIXED REVIEW

SHORT ANSWER Answer the following questions in the space provided.

1. Consider the following data table:

Approximate pressure (kPa)	Altitude above sea level (km)
100	0 (sea level)
50	5.5 (peak of Mt. Kilimanjaro)
25	11 (jet cruising altitude)
< 0.1	22 (ozone layer)

   a. Explain briefly why the pressure decreases as the altitude increases.

   As the altitude increases, there are fewer gas molecules above; therefore, there are fewer gas molecules to exert their pressure.

   b. A few places on Earth are below sea level (the Dead Sea, for example). What would be true about the average atmospheric pressure there?

   It would exceed 100 kPa at places below sea level.

2. Explain how the ideal gas law can be simplified to give Avogadro's law, expressed as $\frac{V}{n} = k$, when the pressure and temperature of a gas are held constant.

   Rearrange $PV = nRT$ to obtain $\frac{V}{n} = \frac{RT}{P}$. Because every value for $\frac{RT}{P}$ is the same, its overall value is constant; therefore, $\frac{V}{n} = k$.

PROBLEMS Write the answer on the line to the left. Show all your work in the space provided.

3. Convert a pressure of 0.400 atm to the following units:

   _____304_____ a. torr

   _____4.05×10^4_____ b. Pa

MIXED REVIEW continued

4. _____226 mL_____ A 250. mL sample of gas is collected at 57°C. What volume will the gas sample occupy at 25°C?

5. _____0.7 L_____ H_2 reacts according to the following equation representing the synthesis of ammonia gas:

$$N_2(g) + 3H_2(g) \rightarrow 2NH_3(g)$$

If 1 L of H_2 is consumed, what volume of ammonia will be produced at constant temperature and pressure, based on Gay-Lussac's law of combining volumes?

6. _____3.15×10^3 kPa_____ A 7.00 L sample of argon gas at 420. K exerts a pressure of 625 kPa. If the gas is compressed to 1.25 L and the temperature is lowered to 350. K, what will be its new pressure?

7. _____2.1×10^3 L_____ Chlorine in the upper atmosphere can destroy ozone molecules, O_3. The reaction can be represented by the following equation:

$$Cl_2(g) + 2O_3(g) \rightarrow 2ClO(g) + 2O_2(g)$$

How many liters of ozone can be destroyed at 220. K and 5.0 kPa if 200.0 g of chlorine gas react with it?

8. _____32 g/mol_____ A gas of unknown molar mass is observed to effuse through a small hole at one-fourth the effusion rate of hydrogen. Estimate the molar mass of this gas. (Round the molar mass of hydrogen to two significant figures.)

Name _____ Date _____ Class _____

CHAPTER 12 REVIEW
Solutions

SECTION 1

SHORT ANSWER Answer the following questions in the space provided.

1. Match the type of mixture on the left to its representative particle diameter on the right.

    __c__ solutions   (a) larger than 1000 nm

    __a__ suspensions   (b) 1 nm to 1000 nm

    __b__ colloids   (c) smaller than 1 nm

2. Identify the solvent in each of the following examples:

    __alcohol__ a. tincture of iodine (iodine dissolved in ethyl alcohol)

    __water__ b. sea water

    __the gels__ c. water-absorbing super gels

3. A certain mixture has the following properties:
    - No solid settles out during a 48-hour period.
    - The path of a flashlight beam is easily seen through the mixture.
    - It appears to be homogeneous under a hand lens but not under a microscope.

    Is the mixture a suspension, colloid, or true solution? Explain your answer.

    The mixture is a colloid. The properties are consistent with those reported in Table 3 on page 404 of the text. The particle size is small, but not too small, and the mixture exhibits the Tyndall effect.

4. Define each of the following terms:

    a. alloy

    a homogeneous mixture of two or more solid metals

    b. electrolyte

    a substance that dissolves in water to form a solution that conducts an electric current

MODERN CHEMISTRY   SOLUTIONS

SECTION 1 continued

　　c. aerosol

　　a colloidal dispersion of a solid or a liquid in a gas

　　d. aqueous solution

　　a mixture with a soluble solute and water as the solvent

5. For each of the following types of solutions, give an example other than those listed in **Table 1** on page 402 of the text:

　　a. a gas in a liquid

　　oxygen gas dissolved in water (needed by fish)

　　b. a liquid in a liquid

　　antifreeze, which is ethylene glycol dissolved in water

　　c. a solid in a liquid

　　salt dissolved in water or iodine in alcohol

6. Using the following models of solutions shown at the particle level, indicate which will conduct electricity. Give a reason for each model.

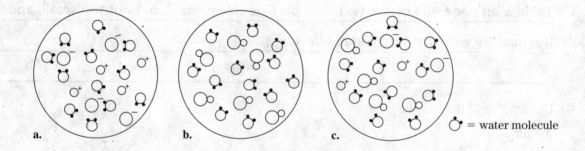

　　a. will conduct electricity because ions are present
　　b. will not conduct electricity because ions are not present
　　c. will conduct electricity slightly because some ions have formed

CHAPTER 12 REVIEW

Solutions

SECTION 2

SHORT ANSWER Answer the following questions in the space provided.

1. The following are statements about the dissolving process. Explain each one at the molecular level.

   a. Increasing the pressure of a solute gas above a liquid solution increases the solubility of the gas in the liquid.

   Increasing the pressure of the solute gas above the solution puts stress on the equilibrium of the system. Gas molecules collide with the liquid surface more often, causing an increase in the rate of gas molecules entering into solution.

   b. Increasing the temperature of water speeds up the rate at which many solids dissolve in this solvent.

   As the temperature of the water increases, water molecules move faster, increasing their average kinetic energy. At higher temperatures, collisions between the water molecules and the solute are more frequent and are of higher energy than at lower temperatures. This helps to separate solute particles from one another and to disperse them among the water molecules.

   c. Increasing the surface area of a solid solute speeds up the rate at which it dissolves in a liquid solvent.

   Increasing the surface area of a solid exposes more of the solute to the solvent, allowing the solvent to come into contact with more of the solute in a shorter length of time.

2. The solubility of $KClO_3$ at 25°C is 10. g of solute per 100. g of H_2O.

   a. If 15 g of $KClO_3$ are stirred into 100 g of water at 25°C, how much of the $KClO_3$ will dissolve? Is the solution saturated, unsaturated, or supersaturated?

   10 g of $KClO_3$ will dissolve, but 5 g will not, despite thorough stirring. The solution is saturated.

SECTION 2 continued

b. If 15 g of $KClO_3$ are stirred into 200 g of water at 25°C, how much of the $KClO_3$ will dissolve? Is the solution saturated, unsaturated, or supersaturated?

All 15 g of $KClO_3$ will dissolve; the solution is unsaturated.

PROBLEMS Write the answer on the line to the left. Show all your work in the space provided.

3. Use the data in **Table 4** on page 410 of the text to answer the following questions:

   __250. g__ **a.** How many grams of LiCl are needed to make a saturated solution with 300. g of water at 20°C?

   __50. g__ **b.** What is the minimum amount of water needed to dissolve 51 g of $NaNO_3$ at 40°C?

   __KI__ **c.** Which solute forms a saturated solution when 36 g of it are dissolved in 25 g of water at 20°C?

4. KOH is an ionic solid readily soluble in water.

   __−1.027 kJ/g__ **a.** What is its enthalpy of solution in kJ/g? Refer to the data in **Table 5** on page 416 of the text.

   b. Will the temperature of the system increase or decrease as the dissolution of KOH proceeds? Why?

   The temperature of the system will increase because the enthalpy of solution is

   negative, indicating that the reaction is exothermic, giving off energy as heat and

   warming up the system.

CHAPTER 12 REVIEW

Solutions

SECTION 3

SHORT ANSWER Answer the following questions in the space provided.

1. Describe the errors made by the following students in making molar solutions.

   a. James needs a 0.600 M solution of KCl. He measures out 0.600 g of KCl and adds 1 L of water to the solid.

   James made several errors. First, 0.600 mol of KCl does not have a mass of 0.600 g.

   Also, adding 1.0 L of water to the solid does not produce 1.0 L of solution. He did

   not make a 0.600 M solution.

   b. Mary needs a 0.02 M solution of $NaNO_3$. She calculates that she needs 2.00 g of $NaNO_3$ for 0.02 mol. She puts this solid into a 1.00 L volumetric flask and fills the flask to the 1.00 L mark.

   Mary did not produce the required solution either. First, 0.02 mol of $NaNO_3$ has a

   mass of 1.70 g, not 2.00 g. Also, she should have made sure the solute was

   completely dissolved before continuing to fill the volumetric flask to the 1.00 L mark.

PROBLEMS Write the answer on the line to the left. Show all of your work in the space provided.

2. _____0.33 M_____ What is the molarity of a solution made by dissolving 2.0 mol of solute in 6.0 L of solvent?

3. _____1.0 m_____ CH_3OH is soluble in water. What is the molality of a solution made by dissolving 8.0 g of CH_3OH in 250. g of water?

MODERN CHEMISTRY   SOLUTIONS **107**

SECTION 3 continued

4. Marble chips effervesce when treated with hydrochloric acid. This reaction is represented by the following equation:

$$CaCO_3(s) + 2HCl(aq) \rightarrow CaCl_2(aq) + CO_2(g) + H_2O(l)$$

To produce a reaction, 25.0 mL of 4.0 M HCl is added to excess $CaCO_3$.

__0.10 mol__ a. How many moles of HCl are consumed in this reaction?

__1.1 L__ b. How many liters of CO_2 are produced at STP?

__5.0 g__ c. How many grams of $CaCO_3$ are consumed?

5. Tincture of iodine is $I_2(s)$ dissolved in ethanol, C_2H_5OH. A 1% solution of tincture of iodine is 10.0 g of solute for 1000. g of solution.

__990. g__ a. How many grams of solvent are present in 1000. g of this solution?

__0.0394 mol__ b. How many moles of solute are in 10.0 g of I_2?

__0.0398 m__ c. What is the molality of this 1% solution?

d. To determine a solution's molarity, the density of that solution can be used. Explain how you would use the density of the tincture of iodine solution to calculate its molarity.

The density of a solution can be expressed in g/mL or in kg/L. Divide 1.00 kg by

the solution's density to find the volume of solution in liters. Then divide 0.0394 mol

by this volume to arrive at the molarity.

CHAPTER 12 REVIEW

Solutions

MIXED REVIEW

SHORT ANSWER Answer the following questions in the space provided.

1. Solid $CaCl_2$ does not conduct electricity. Explain why it is considered to be an electrolyte.

   $CaCl_2$ is an ionic solid. In the crystal form, its ions are locked in position. Dissolving the crystal in water releases the ions to move freely, allowing them to conduct electricity.

2. Explain the following statements at the molecular level:

   a. Generally, a polar liquid and a nonpolar liquid are immiscible.

   Polar molecules tend to attract one another, forcing the nonpolar molecules to remain in a separate layer.

   b. Carbonated soft drinks taste flat when they warm up.

   The solubility of gases usually decreases as the temperature of the solution increases. At higher temperatures, more CO_2 molecules escape through the liquid's surface, leaving fewer molecules in solution to effervesce.

3. An unknown compound is observed to mix with toluene, $C_6H_5CH_3$, but not with water.

   a. Is the unknown compound ionic, polar covalent, or nonpolar covalent? Explain your answer.

   nonpolar covalent, because it mixes with nonpolar toluene and not with polar water

   b. Suppose the unknown compound is also a liquid. Will it be able to dissolve table salt? Explain why or why not.

   No; being nonpolar, the solvent molecules are unable to remove ions from sodium chloride's crystal surfaces.

MIXED REVIEW continued

PROBLEMS Write the answer on the line to the left. Show all your work in the space provided.

4. Consider 500. mL of a 0.30 M $CuSO_4$ solution.

   __0.15 mol__ a. How many moles of solute are present in this solution?

   __24 g__ b. How many grams of solute were used to prepare this solution?

5. a. If a solution is electrically neutral, can all of its ions have the same type of charge? Explain your answer.
   No; to be neutral the total positive charge must equal the total negative charge.

   __6.0×10^{13} ions__ b. The concentration of the OH^- ions in pure water is known to be 1.0×10^{-7} M. How many OH^- ions are present in each milliliter of pure water?

6. 90. g of $CaBr_2$ are dissolved in 900. g of water.

   __900. mL__ a. What volume does the 900. g of water occupy if its density is 1.00 g/mL?

   __0.50 m__ b. What is the molality of this solution?

CHAPTER 13 REVIEW
Ions in Aqueous Solutions and Colligative Properties

SECTION 1

SHORT ANSWER Answer the following questions in the space provided.

1. Use the guidelines in **Table 1** on page 437 of the text to predict the solubility of the following compounds in water:

    __soluble__ a. magnesium nitrate

    __insoluble__ b. barium sulfate

    __insoluble__ c. calcium carbonate

    __soluble__ d. ammonium phosphate

2. 1.0 mol of magnesium acetate is dissolved in water.

    __$Mg(CH_3COO)_2$__ a. Write the formula for magnesium acetate.

    __3.0 mol__ b. How many moles of ions are released into solution?

    __0.60 mol__ c. How many moles of ions are released into a solution made from 0.20 mol magnesium acetate dissolved in water?

3. Write the formula for the precipitate formed

    __$Mg_3(PO_4)_2$__ a. when solutions of magnesium chloride and potassium phosphate are combined.

    __Ag_2S__ b. when solutions of sodium sulfide and silver nitrate are combined.

4. Write ionic equations for the dissolution of the following compounds:

    a. $Na_3PO_4(s)$

    $Na_3PO_4(s) \rightarrow 3Na^+(aq) + PO_4^{3-}(aq)$

    b. iron(III) sulfate(s)

    $Fe_2(SO_4)_3(s) \rightarrow 2Fe^{3+}(aq) + 3SO_4^{2-}(aq)$

5. a. Write the net ionic equation for the reaction that occurs when solutions of lead(II) nitrate and ammonium sulfate are combined.

    $Pb^{2+}(aq) + SO_4^{2-}(aq) \rightarrow PbSO_4(s)$

    b. What are the spectator ions in this system?

    NO_3^- and NH_4^+ are spectator ions.

SECTION 1 continued

6. The following solutions are combined in a beaker: NaCl, Na_3PO_4, and $Ba(NO_3)_2$.

   a. Will a precipitate form when the above solutions are combined? If so, write the name and formula of the precipitate.

   Yes; barium phosphate, $Ba_3(PO_4)_2$, forms as a precipitate.

   b. List all spectator ions present in this system.

   Na^+, Cl^-, and NO_3^- are spectator ions in this system.

7. It is possible to have spectator ions present in many chemical systems, not just in precipitation reactions. Consider this example:

$$Al(s) + HCl(aq) \rightarrow AlCl_3(aq) + H_2(g) \text{ (unbalanced)}$$

   __True__ **a.** In an aqueous solution of HCl, virtually every HCl molecule is ionized. True or False?

   __$Cl^-(aq)$__ **b.** There is only one spectator ion in this system. Is it $Al^{3+}(aq)$, $H^+(aq)$, or $Cl^-(aq)$?

   c. Balance the above equation.

   $2Al(s) + 6HCl(aq) \rightarrow 2AlCl_3(aq) + 3H_2(g)$

   __11 L__ **d.** If 9.0 g of Al metal react with excess HCl according to the balanced equation in part **c**, what volume of hydrogen gas at STP will be produced? Show all your work.

8. Acetic acid, CH_3CO_2H, is a weak electrolyte. Write an equation to represent its ionization in water. Include the hydronium ion, H_3O^+.

   $CH_3CO_2H(aq) + H_2O(l) \rightleftharpoons CH_3CO_2^-(aq) + H_3O^+(aq)$

CHAPTER 13 REVIEW

Ions in Aqueous Solutions and Colligative Properties

SECTION 2

PROBLEMS Write the answer on the line to the left. Show all your work in the space provided.

1. __100.102°C__ a. Predict the boiling point of a 0.200 m solution of glucose in water.

   __100.204°C__ b. Predict the boiling point of a 0.200 m solution of potassium iodide in water.

2. A chief ingredient of antifreeze is liquid ethylene glycol, $C_2H_4(OH)_2$. Assume $C_2H_4(OH)_2$ is added to a car radiator that is holding 5.0 kg of water.

   __48 mol__ a. How many moles of ethylene glycol should be added to the radiator to lower the freezing point of the water from 0°C to −18°C?

   __3.0 × 10³ g__ b. How many grams of ethylene glycol does the quantity in part **a** represent?

   __2.7 L__ c. Ethylene glycol has a density of 1.1 kg/L. How many liters of $C_2H_4(OH)_2$ should be added to the water in the radiator to prevent freezing down to −18°C?

SECTION 2 continued

    d. In World War II, soldiers in the Sahara Desert needed a supply of antifreeze to protect the radiators of their vehicles. The temperature in the Sahara almost never drops to 0°C, so why was the antifreeze necessary?

Antifreeze also raises the boiling point of water. It was needed to help prevent the water in the radiators of the vehicles from boiling over.

3. An important use of colligative properties is to determine the molar mass of unknown substances. The following situation is an example: 12.0 g of unknown compound X, a nonpolar nonelectrolyte, is dissolved in 100.0 g of melted camphor. The resulting solution freezes at 99.4°C. Consult **Table 2** on page 448 of the text for any other data needed to answer the following questions:

    ___79.4°C___   **a.** By how many °C did the freezing point of camphor change from its normal freezing point?

    ___2.00 m___   **b.** What is the molality of the solution of camphor and compound X, based on freezing-point data?

    ___120. g___   **c.** If there are 12.0 g of compound X per 100.0 g of camphor, how many grams of compound X are there per kilogram of camphor?

    ___60.0 g/mol___   **d.** What is the molar mass of compound X?

4. Explain why the ability of a solution to conduct an electric current is not a colligative property.

Electrical conductivity depends on the nature of the solute, unlike colligative properties, which depend only on concentration of solute particles.

CHAPTER 13 REVIEW
Ions in Aqueous Solutions and Colligative Properties

MIXED REVIEW

SHORT ANSWER Answer the following questions in the space provided.

1. Match the four compounds on the right to their descriptions on the left.
   - __b__ an ionic compound that is quite soluble in water — (a) HCl
   - __c__ an ionic compound that is not very soluble in water — (b) $NaNO_3$
   - __a__ a molecular compound that ionizes in water — (c) AgCl
   - __d__ a molecular compound that does not ionize in water — (d) C_2H_5OH

2. Consider nonvolatile nonelectrolytes dissolved in various liquid solvents to complete the following statements:

   __solute__ a. The change in the boiling point does *not* vary with the identity of the ___ (solute, solvent), assuming all other factors remain constant.

   __solvent__ b. The change in the boiling point varies with the identity of the ___ (solute, solvent), assuming all other factors remain constant.

   __increases__ c. The change in the boiling point becomes greater as the concentration of the solute in solution ___ (increases, decreases).

3. a. Name two compounds in solution that could be combined to cause the formation of a calcium carbonate precipitate.

   Answers will vary; any soluble calcium salt mixed with any soluble carbonate will form the precipitate. One example is calcium nitrate with sodium carbonate.

   b. Identify any spectator ions in the system you described in part **a**.

   In the example given, sodium and nitrate ions are spectator ions.

   c. Write the net ionic equation for the formation of calcium carbonate.

   $Ca^{2+}(aq) + CO_3^{2-}(aq) \rightarrow CaCO_3(s)$

4. Explain why applying rock salt (impure NaCl) to an icy sidewalk hastens the melting process.

   The vapor pressure of the NaCl solution that forms is lower than the vapor pressure of pure water at 0°C. The lower vapor pressure of the NaCl solution results in a lower freezing point.

Name _____ Date _____ Class _____

MIXED REVIEW continued

PROBLEMS Write the answer on the line to the left. Show all your work in the space provided.

5. __**13.4 m**__ Some insects survive cold winters by generating an antifreeze inside their cells. The antifreeze produced is glycerol, $C_3H_5(OH)_3$, a nonvolatile nonelectrolyte that is quite soluble in water. What must the molality of a glycerol solution be to lower the freezing point of water to $-25.0°C$?

6. __**2.14 g**__ How many grams of methanol, CH_3OH, should be added to 200. g of acetic acid to lower its freezing point by $1.30°C$? Refer to **Table 2** on page 448 of the text for any necessary data.

7. __**0.67 m**__ The boiling point of a solution of glucose, $C_6H_{12}O_6$, and water was recorded to be $100.34°C$. Calculate the molality of this solution.

8. HF(*aq*) is a weak acid. A 0.05 mol sample of HF is added to 1.0 kg of water.
   a. Write the equation for the ionization of HF to form hydronium ions.
   HF(*aq*) + H$_2$O(*l*) → H$_3$O$^+$(*aq*) + F$^-$(*aq*)

   __**0.10 mol**__ b. If HF became 100% ionized, how many moles of its ions would be released?

9. __**c**__ Which solution has the highest osmotic pressure?
   a. 0.1 *m* glucose
   b. 0.1 *m* sucrose
   c. 0.5 *m* glucose
   d. 0.2 *m* sucrose

CHAPTER 14 REVIEW
Acids and Bases

SECTION 1

SHORT ANSWER Answer the following questions in the space provided.

1. Name the following compounds as acids:

    _____sulfuric acid_____ a. H_2SO_4

    _____sulfurous acid_____ b. H_2SO_3

    _____hydrosulfuric acid_____ c. H_2S

    _____perchloric acid_____ d. $HClO_4$

    _____hydrocyanic acid_____ e. hydrogen cyanide

2. _____H_2S_____ Which (if any) of the acids mentioned in item **1** are binary acids?

3. Write formulas for the following acids:

    _____HNO_2_____ a. nitrous acid

    _____HBr_____ b. hydrobromic acid

    _____H_3PO_4_____ c. phosphoric acid

    _____CH_3COOH_____ d. acetic acid

    _____HClO_____ e. hypochlorous acid

4. Calcium selenate has the formula $CaSeO_4$.

    _____H_2SeO_4_____ a. What is the formula for selenic acid?

    _____H_2SeO_3_____ b. What is the formula for selenous acid?

5. Use an activity series to identify two metals that will not generate hydrogen gas when treated with an acid.

    Choose from Cu, Ag, Au, Pt, Pd, or Hg.

6. Write balanced chemical equations for the following reactions of acids and bases:

    a. aluminum metal with dilute nitric acid

    $2Al(s) + 6HNO_3(aq) \rightarrow 2Al(NO_3)_3(aq) + 3H_2(g)$

    b. calcium hydroxide solution with acetic acid

    $Ca(OH)_2(aq) + 2CH_3COOH(aq) \rightarrow Ca(CH_3COO)_2(aq) + 2H_2O(l)$

SECTION 1 continued

7. Write net ionic equations that represent the following reactions:

   a. the ionization of $HClO_3$ in water

   $HClO_3(aq) + H_2O(l) \rightarrow H_3O^+(aq) + ClO_3^-(aq)$

   b. NH_3 functioning as an Arrhenius base

   $NH_3(aq) + H_2O(l) \rightleftarrows NH_4^+(aq) + OH^-(aq)$

8. a. Explain how strong acid solutions carry an electric current.

   Strong acids ionize completely in solution. These ions are free to move, making it possible for an electric current to pass through the solution.

   b. Will a strong acid or a weak acid conduct electricity better, assuming all other factors remain constant? Explain why one is a better conductor.

   Strong acids conduct electricity better because they have many more ions present per liter of solution than do weak acids of the same concentration.

9. Most acids react with solid carbonates, as in the following equation:

   $CaCO_3(s) + HCl(aq) \rightarrow CaCl_2(aq) + H_2O(l) + CO_2(g)$ (unbalanced)

   a. Balance the above equation.

   $CaCO_3(s) + 2HCl(aq) \rightarrow CaCl_2(aq) + H_2O(l) + CO_2(g)$

   b. Write the net ionic equation for the above reaction.

   $CO_3^{2-}(aq) + 2H^+(aq) \rightarrow H_2O(l) + CO_2(g)$

   __Ca²⁺ and Cl⁻__ c. Identify all spectator ions in this system.

   __1.1 L__ d. How many liters of CO_2 form at STP if 5.0 g of $CaCO_3$ are treated with excess hydrochloric acid? Show all your work.

CHAPTER 14 REVIEW
Acids and Bases

SECTION 2

SHORT ANSWER Answer the following questions in the space provided.

1. a. Write the two equations that show the two-stage ionization of sulfurous acid in water.

    stage 1: $H_2SO_3(aq) + H_2O(l) \rightleftharpoons H_3O^+(aq) + HSO_3^-(aq)$

    stage 2: $HSO_3^-(aq) + H_2O(l) \rightleftharpoons H_3O^+(aq) + SO_3^{2-}(aq)$

    b. Which stage of ionization usually produces more ions? Explain your answer.

    Stage 1; for most polyprotic acids, the concentration of ions formed in the first ionization is the greatest.

2. a. Define a Lewis base. Can OH^- function as a Lewis base? Explain your answer.

    A Lewis base is a species that donates an electron pair to form a covalent bond. Yes, OH^- is a Lewis base. It has an electron pair available to donate. For example,

    $OH^-(aq) + H^+(aq) \rightarrow H_2O(l)$.

    b. Define a Lewis acid. Can H^+ function as a Lewis acid? Explain your answer.

    A Lewis acid is a species that accepts an electron pair to form a covalent bond. H^+ is a Lewis acid because it can accept an electron pair from a base. For example,

    $H^+(aq) + OH^-(aq) \rightarrow H_2O(l)$.

3. Identify the Brønsted-Lowry acid and the Brønsted-Lowry base on the reactant side of each of the following equations for reactions that occur in aqueous solution. Explain your answers.

    a. $H_2O(l) + HNO_3(aq) \rightarrow H_3O^+(aq) + NO_3^-(aq)$

    HNO_3 is the Brønsted-Lowry acid because it donates a proton to the H_2O. The H_2O is the Brønsted-Lowry base because it is the proton acceptor.

    b. $HF(aq) + HS^-(aq) \rightarrow H_2S(aq) + F^-(aq)$

    HF is the Brønsted-Lowry acid because it donates a proton to the HS^-. The HS^- is the Brønsted-Lowry base because it is the proton acceptor.

SECTION 2 continued

4. a. Write the equation for the first ionization of H_2CO_3 in aqueous solution. Assume that water serves as the reactant that attaches to the hydrogen ion released from the H_2CO_3. Which of the reactants is the Brønsted-Lowry acid, and which is the Brønsted-Lowry base? Explain your answer.

   $H_2CO_3(aq) + H_2O(l) \rightleftarrows HCO_3^-(aq) + H_3O^+(aq)$. The H_2CO_3 is the Brønsted-Lowry acid because it donates a proton to the H_2O. The H_2O is the Brønsted-Lowry base because it accepts the proton.

   b. Write the equation for the second ionization, that of the ion that was formed by the H_2CO_3 in the reaction you described above. Again, assume that water serves as the reactant that attaches to the hydrogen ion released. Which of the reactants is the Brønsted-Lowry acid, and which is the Brønsted-Lowry base? Explain your answer.

   $HCO_3^-(aq) + H_2O(l) \rightarrow CO_3^{2-}(aq) + H_3O^+(aq)$. The HCO_3^- is the Brønsted-Lowry acid because it donates a proton to the H_2O. The H_2O is the Brønsted-Lowry base because it accepts the proton.

   c. What is the name for a substance, such as H_2CO_3, that can donate two protons?

   a diprotic acid

5. a. How many electron pairs surround an atom of boron (B, element 5) bonded in the compound BCl_3?

   three

   b. How many electron pairs surround an atom of nitrogen (N, element 7) in the compound NF_3?

   four

   c. Write an equation for the reaction between the two compounds above. Assume that they react in a 1:1 ratio to form one molecule as product.

   $BCl_3 + NF_3 \rightarrow BCl_3NF_3$

   d. Assuming that the B and the N are covalently bonded to each other in the product, which of the reactants is the Lewis acid? Is this reactant also a Brønsted-Lowry acid? Explain your answers.

   BCl_3 is the Lewis acid because it accepts an electron pair in forming a covalent bond. It is not a Brønsted-Lowry acid, because it is not donating a proton.

   e. Which of the reactants is the Lewis base? Explain your answer.

   NF_3 is the Lewis base because it donates an electron pair in forming a covalent bond.

CHAPTER 14 REVIEW
Acids and Bases

SECTION 3

SHORT ANSWER Answer the following questions in the space provided.

1. Answer the following questions according to the Brønsted-Lowry definitions of acids and bases:

    __HSO_3^-__ a. What is the conjugate base of H_2SO_3?

    __NH_3__ b. What is the conjugate base of NH_4^+?

    __OH^-__ c. What is the conjugate base of H_2O?

    __H_3O^+__ d. What is the conjugate acid of H_2O?

    __$H_2AsO_4^-$__ e. What is the conjugate acid of $HAsO_4^{2-}$?

2. Consider the reaction described by the following equation:

    $$NH_4^+(aq) + CO_3^{2-}(aq) \rightleftharpoons NH_3(aq) + HCO_3^-(aq)$$

    a. If NH_4^+ is considered acid 1, identify the other three terms as acid 2, base 1, and base 2 to indicate the conjugate acid-base pairs.

    __base 2__ CO_3^{2-}

    __acid 2__ HCO_3^-

    __base 1__ NH_3

    __True__ b. A proton has been transferred from acid 1 to base 2 in the above reaction. True or False?

3. Consider the neutralization reaction described by the equation: $HCO_3^-(aq) + OH^-(aq) \rightleftharpoons CO_3^{2-}(aq) + H_2O(l)$

    a. Label the conjugate acid-base pairs in this system.

    $HCO_3^-(aq) + OH^-(aq) \rightleftharpoons CO_3^{2-}(aq) + H_2O(l)$

    acid 1   base 2   base 1   acid 2

    b. Is the forward or reverse reaction favored? Explain your answer.

    The forward reaction is favored. The weaker acid and weaker base are produced in the forward reaction. HCO_3^- competes more strongly with H_2O to donate a proton, and OH^- competes more strongly with CO_3^{2-} to acquire a proton, causing the forward reaction to be favored.

SECTION 3 continued

4. Table 6 on page 485 of the text lists several amphoteric species, but only one other than water is neutral.

___NH₃___ **a.** Identify that neutral compound.

b. Write two equations that demonstrate this compound's amphoteric properties.

(Answers will vary, but one equation should form NH_4^+, the other NH_2^-.)

$NH_3(aq) + HCl(aq) \rightarrow NH_4^+(aq) + Cl^-(aq)$

$NH_3(aq) + H^-(aq) \rightarrow NH_2^-(aq) + H_2(g)$

5. Write the formula for the salt formed in each of the following neutralization reactions:

___K_3PO_4___ **a.** potassium hydroxide combined with phosphoric acid

___$Ca(NO_2)_2$___ **b.** calcium hydroxide combined with nitrous acid

___$BaBr_2$___ **c.** hydrobromic acid combined with barium hydroxide

___Li_2SO_4___ **d.** lithium hydroxide combined with sulfuric acid

6. Consider the following unbalanced equation for a neutralization reaction:

$$H_2SO_4(aq) + NaOH(aq) \rightarrow Na_2SO_4(aq) + H_2O(l)$$

a. Balance the equation.

$H_2SO_4(aq) + 2NaOH(aq) \rightarrow Na_2SO_4(aq) + 2H_2O(l)$

___Na^+ and SO_4^{2-}___ **b.** In this system there are two spectator ions. Identify them.

___1:2___ **c.** For the reaction to completely consume all reactants, what should be the mole ratio of acid to base?

7. The gases that produce acid rain are often referred to as NO_x and SO_x.

a. List three specific examples of these gases.

Some examples are NO, NO_2, SO_2, and SO_3.

b. Coal- and oil-burning power plants oxidize any sulfur in their fuel as it burns in air, and this forms SO_2 gas. The SO_2 is further oxidized by O_2 in our atmosphere, forming SO_3 gas. The SO_3 gas can combine with water to form sulfuric acid. Write balanced chemical equations to illustrate these three reactions.

$S(s) + O_2(g) \rightarrow SO_2(g)$

$2SO_2(g) + O_2(g) \rightarrow 2SO_3(g)$

$SO_3(g) + H_2O(l) \rightarrow H_2SO_4(aq)$

c. Industrial plants making fertilizers and detergents release nitrogen oxide gases into the air. Write a balanced equation for converting $N_2O_5(g)$ into nitric acid by reacting it with water.

$N_2O_5(g) + H_2O(l) \rightarrow 2HNO_3(aq)$

CHAPTER 14 REVIEW
Acids and Bases

MIXED REVIEW

SHORT ANSWER Answer the following questions in the space provided.

1.
   a. Write the formula for hypochlorous acid. — **HClO**
   b. Write the name for HF(aq). — **hydrofluoric acid**
   c. If $Pb(C_2O_4)_2$ is lead(IV) oxalate, what is the formula for oxalic acid? — **$H_2C_2O_4$**
   d. Name the acid that is present in vinegar. — **acetic acid**

2. Answer the following questions according to the Brønsted-Lowry acid-base theory. Consult **Table 6** on page 485 of the text as needed.
   a. What is the conjugate base of H_2S? — **HS^-**
   b. What is the conjugate base of HPO_4^{2-}? — **PO_4^{3-}**
   c. What is the conjugate acid of NH_3? — **NH_4^+**

3. Consider the reaction represented by the following equation:

   $$OH^-(aq) + HCO_3^-(aq) \rightarrow H_2O(l) + CO_3^{2-}(aq)$$

   If OH^- is considered base 1, what are acid 1, acid 2, and base 2?
   a. acid 1 — **H_2O**
   b. acid 2 — **HCO_3^-**
   c. base 2 — **CO_3^{2-}**

4. Write the formula for the salt that is produced in each of the following neutralization reactions:
   a. sulfurous acid combined with potassium hydroxide — **K_2SO_3**
   b. calcium hydroxide combined with phosphoric acid — **$Ca_3(PO_4)_2$**

5. Carbonic acid releases H_3O^+ ions into water in two stages.
   a. Write equations representing each stage.
   stage 1: $H_2CO_3(aq) + H_2O(l) \rightleftarrows H_3O^+(aq) + HCO_3^-(aq)$
   stage 2: $HCO_3^-(aq) + H_2O(l) \rightleftarrows H_3O^+(aq) + CO_3^{2-}(aq)$

   b. Which stage releases more ions into solution? — **stage 1**

MIXED REVIEW continued

6. Glacial acetic acid is a highly viscous liquid that is close to 100% CH₃COOH. When it mixes with water, it forms dilute acetic acid.

   a. When making a dilute acid solution, should you add acid to water or water to acid? Explain your answer.

   Add acid to water to achieve a thorough mixing of a denser acid with a slow release of heat and to avoid splashing concentrated acid.

   b. Glacial acetic acid does not conduct electricity, but dilute acetic acid does. Explain this statement.

   Glacial acetic acid exists as neutral molecules. In the presence of water, some of those molecules ionize into H^+ and CH_3COO^-, which cause the solution to conduct electricity.

   c. Dilute acetic acid does not conduct electricity as well as dilute nitric acid at the same concentration. Is acetic acid a strong or weak acid?

   weak

   d. Although there are four H atoms per molecule, acetic acid is monoprotic. Show the structural formula for CH₃COOH, and indicate the H atom that ionizes.

   [Structural formula of acetic acid shown]   **The H bonded to the O ionizes.**

   e. _____**30. g**_____ How many grams of glacial acetic acid should be used to make 250 mL of 2.00 M acetic acid? Show all your work.

7. The overall effect of acid rain on lakes and ponds is partially determined by the geology of the lake bed. In some cases, the rock is limestone, which is rich in calcium carbonate. Calcium carbonate reacts with the acid in lake water according to the following (incomplete) ionic equation:

   $$CaCO_3(s) + 2H_3O^+(aq) \rightarrow$$

   a. Complete the ionic equation begun above.

   $CaCO_3(s) + 2H_3O^+(aq) \rightarrow Ca^{2+}(aq) + CO_2(g) + 3H_2O(l)$

   b. If this reaction is the only reaction involving H_3O^+ occurring in the lake, does the concentration of H_3O^+ in the lake water increase or decrease? What effect does this have on the acidity of the lake water?

   It decreases, making the lake water less acidic.

CHAPTER 15 REVIEW
Acid-Base Titration and pH

SECTION 1

SHORT ANSWER Answer the following questions in the space provided.

1. Calculate the following values without using a calculator.

   1×10^{-8} M **a.** The $[H_3O^+]$ is 1×10^{-6} M in a solution. Calculate the $[OH^-]$.

   1×10^{-5} M **b.** The $[H_3O^+]$ is 1×10^{-9} M in a solution. Calculate the $[OH^-]$.

   1×10^{-2} M **c.** The $[OH^-]$ is 1×10^{-12} M in a solution. Calculate the $[H_3O^+]$.

   2×10^{-2} M **d.** The $[OH^-]$ in part **c** is reduced by half, to 0.5×10^{-12} M. Calculate the $[H_3O^+]$.

   inversely **e.** The $[H_3O^+]$ and $[OH^-]$ are ____ (directly, inversely, or not) proportional in any system involving water.

2. Calculate the following values without using a calculator.

   12.0 **a.** The pH of a solution is 2.0. Calculate the pOH.

   9.27 **b.** The pOH of a solution is 4.73. Calculate the pH.

   3.0 **c.** The $[H_3O^+]$ in a solution is 1×10^{-3} M. Calculate the pH.

   1×10^{-5} M **d.** The pOH of a solution is 5.0. Calculate the $[OH^-]$.

   1×10^{-13} M **e.** The pH of a solution is 1.0. Calculate the $[OH^-]$.

3. Calculate the following values.

   4.631 **a.** The $[H_3O^+]$ is 2.34×10^{-5} M in a solution. Calculate the pH.

   3.16×10^{-4} M **b.** The pOH of a solution is 3.5. Calculate the $[OH^-]$.

   2.2×10^{-7} M **c.** The $[H_3O^+]$ is 4.6×10^{-8} M in a solution. Calculate the $[OH^-]$.

PROBLEMS Write the answer on the line to the left. Show all your work in the space provided.

4. $[H_3O^+]$ in an aqueous solution = 2.3×10^{-3} M.

   4.3×10^{-12} M **a.** Calculate $[OH^-]$ in this solution.

MODERN CHEMISTRY   ACID-BASE TITRATION AND pH   125

SECTION 1 continued

_____2.64_____ **b.** Calculate the pH of this solution.

_____11.36_____ **c.** Calculate the pOH of this solution.

d. Is the solution acidic, basic, or neutral? Explain your answer.

acidic because the pH is less than 7.0

5. Consider a dilute solution of 0.025 M $Ba(OH)_2$ in answering the following questions.

a. What is the $[OH^-]$ in this solution? Explain your answer.

0.050 M; in solution, $Ba(OH)_2$ releases two OH^- ions per molecule, so $[OH^-]$ is two times the $[Ba(OH)_2]$.

_____12.70_____ **b.** What is the pH of this solution?

6. Vinegar purchased in a store may contain 6 g of CH_3COOH per 100 mL of solution.

_____1 M_____ **a.** What is the molarity of the solute?

b. The actual $[H_3O^+]$ in the vinegar solution in part **a** is 4.2×10^{-3} M. In this solution, has more than 1% or less than 1% of the acetic acid ionized? Explain your answer.

less than 1%; 1% of 1 M would equal 1×10^{-2} hydronium ions, but in fact, less than that amount has been produced.

_____weak_____ **c.** Is acetic acid strong or weak, based on the ionization information from part **b**?

_____2.38_____ **d.** What is the pH of this vinegar solution?

CHAPTER 15 REVIEW

Acid-Base Titration and pH

SECTION 2

SHORT ANSWER Answer the following questions in the space provided.

1. Below is a pH curve from an acid-base titration. On it are labeled three points: X, Y, and Z.

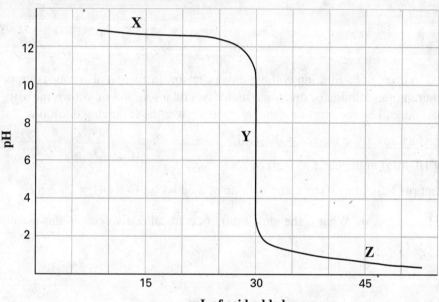

_____Y_____ a. Which point represents the equivalence point?

_____Z_____ b. At which point is there excess acid in the system?

_____X_____ c. At which point is there excess base in the system?

__7.5×10^{-3} mol__ d. If the base solution is 0.250 M and there is one equivalent of OH$^-$ ions for each mole of base, how many moles of OH$^-$ ions are consumed at the end point of the titration?

PROBLEMS Write the answer on the line to the left. Show all your work in the space provided.

2. A standardized solution of 0.065 M HCl is titrated with a saturated solution of calcium hydroxide to determine its molarity and its solubility. It takes 25.0 mL of the base to neutralize 10.0 mL of the acid.

a. Write the balanced molecular equation for this neutralization reaction.

$Ca(OH)_2(aq) + 2HCl(aq) \rightarrow CaCl_2(aq) + 2H_2O(l)$

SECTION 2 continued

__0.013 M__ b. Determine the molarity of the Ca(OH)$_2$ solution.

__0.96 g/L__ c. Based on your answer to part **b**, calculate the solubility of the base in grams per liter of solution. (Hint: What is the concentration of Ca(OH)$_2$ in the saturated solution?)

3. It is possible to carry out a titration without any indicator. Instead, a pH probe is immersed in a beaker containing the solution of unknown molarity, and a solution of known molarity is slowly added from a buret. Use the titration data below to answer the following questions.

   Volume of KOH(*aq*) in the beaker = 30.0 mL

   Molarity of HCl(*aq*) in the buret = 0.50 M

   At the instant pH falls from 10 to 4, the volume of acid added to KOH = 27.8 mL.

   __1:1__ a. What is the mole ratio of chemical equivalents in this system?

   __0.46 M__ b. Calculate the molarity of the KOH solution, based on the above data.

CHAPTER 15 REVIEW
Acid-Base Titration and pH

MIXED REVIEW

SHORT ANSWER Answer the following questions in the space provided.

1. Calculate the following values without using a calculator.

   __4.0__ a. The $[H_3O^+]$ in a solution is 1×10^{-4} M. Calculate the pH.

   __1×10^{-13} M__ b. The pH of a solution is 13.0. Calculate the $[H_3O^+]$.

   __1×10^{-9} M__ c. The $[OH^-]$ in a solution is 1×10^{-5} M. Calculate the $[H_3O^+]$.

   __9.28__ d. The pH of a solution is 4.72. Calculate the pOH.

   __14.00__ e. The $[OH^-]$ in a solution is 1.0 M. Calculate the pH.

2. Calculate the following values.

   __8.204__ a. The $[H_3O^+]$ in a solution is 6.25×10^{-9} M. Calculate the pH.

   __4.6×10^{-3} M__ b. The pOH of a solution is 2.34. Calculate the $[OH^-]$.

   __3×10^{-4} M__ c. The pH of milk of magnesia is approximately 10.5. Calculate the $[OH^-]$.

PROBLEMS Write the answer on the line to the left. Show all your work in the space provided.

3. A 0.0012 M solution of H_2SO_4 is 100% ionized.

   __0.0024 M__ a. What is the $[H_3O^+]$ in the H_2SO_4 solution?

   __4.2×10^{-12} M__ b. What is the $[OH^-]$ in this solution?

   __2.62__ c. What is the pH of this solution?

MIXED REVIEW continued

4. In a titration, a 25.0 mL sample of 0.150 M HCl is neutralized with 44.45 mL of Ba(OH)$_2$.

   a. Write the balanced molecular equation for this reaction.

   $2HCl(aq) + Ba(OH)_2(aq) \rightarrow BaCl_2(aq) + 2H_2O(l)$

   __0.0422 M__ b. What is the molarity of the base solution?

5. 3.09 g of boric acid, H$_3$BO$_3$, are dissolved in 200 mL of solution.

   __0.250 M__ a. Calculate the molarity of the solution.

   b. H$_3$BO$_3$ ionizes in solution in three stages. Write the equation showing the ionization for each stage. Which stage proceeds furthest to completion?

   stage 1: $H_3BO_3(s) + H_2O(l) \rightleftarrows H_3O^+(aq) + H_2BO_3^-(aq)$

   stage 2: $H_2BO_3^-(aq) + H_2O(l) \rightleftarrows H_3O^+(aq) + HBO_3^{2-}(aq)$

   stage 3: $HBO_3^{2-}(aq) + H_2O(l) \rightleftarrows H_3O^+(aq) + BO_3^{3-}(aq)$

   Stage 1 proceeds furthest to completion.

   __1.3×10^{-5} M__ c. What is the [H$_3$O$^+$] in this boric acid solution if the pH = 4.90?

   __less than 1%__ d. Is the percentage ionization of this H$_3$BO$_3$ solution more than or less than 1%?

CHAPTER 16 REVIEW
Reaction Energy

SECTION 1

SHORT ANSWER Answer the following questions in the space provided.

1. For elements in their standard state, the value of $\triangle H_f^0$ is ___0___.

2. The formation and decomposition of water can be represented by the following thermochemical equations:

$$H_2(g) + \tfrac{1}{2}O_2(g) \rightarrow H_2O(g) + 241.8 \text{ kJ/mol}$$

$$H_2O(l) + 241.8 \text{ kJ/mol} \rightarrow H_2(g) + \tfrac{1}{2}O_2(g)$$

___taken in___ a. Is energy being taken in or is it being released as liquid H_2O decomposes?

___positive___ b. What is the appropriate sign for the enthalpy change in this decomposition reaction?

PROBLEMS Write the answer on the line to the left. Show all your work in the space provided.

3. ___70°C___ If 200. g of water at 20°C absorbs 41 840 J of energy, what will its final temperature be?

4. ___28.9 kJ___ Aluminum has a specific heat of 0.900 J/(g·°C). How much energy in kJ is needed to raise the temperature of a 625 g block of aluminum from 30.7°C to 82.1°C?

5. The products in a reaction have an enthalpy of 458 kJ/mol, and the reactants have an enthalpy of 658 kJ/mol.

   ___−200. kJ/mol___ a. What is the value of $\triangle H$ for this reaction?

SECTION 1 continued

__products__ **b.** Which is the more stable part of this system, the reactants or the products?

6. The enthalpy of combustion of acetylene gas is -1301.1 kJ/mol of C_2H_2.

a. Write the balanced thermochemical equation for the complete combustion of C_2H_2.

$$C_2H_2(g) + \tfrac{5}{2}O_2(g) \rightarrow 2CO_2(g) + H_2O(l) + \text{heat energy}$$

__320 kJ__ **b.** If 0.25 mol of C_2H_2 reacts according to the equation in part **a**, how much energy is released?

__78 g__ **c.** How many grams of C_2H_2 are needed to react, according to the equation in part **a**, to release 3900 kJ of energy?

7. __−850. kJ/mol__ Determine the ΔH for the reaction between Al and Fe_2O_3, according to the equation $2Al + Fe_2O_3 \rightarrow Al_2O_3 + 2Fe$. The enthalpy of formation of Al_2O_3 is -1676 kJ/mol. For Fe_2O_3 it is -826 kJ/mol.

8. __−196.0 kJ/mol__ Use the data in **Appendix Table A-14** of the text to determine the ΔH for the following equation.

$$2H_2O_2(l) \rightarrow 2H_2O(l) + O_2(g)$$

CHAPTER 16 REVIEW
Reaction Energy

SECTION 2

SHORT ANSWER Answer the following questions in the space provided.

1. For the following examples, state whether the change in entropy favors the forward or reverse reaction:

   __forward reaction__ a. $HCl(l) \rightleftarrows HCl(g)$

   __reverse reaction__ b. $C_6H_{12}O_6(aq) \rightleftarrows C_6H_{12}O_6(s)$

   __forward reaction__ c. $2NH_3(g) \rightleftarrows N_2(g) + 3H_2(g)$

   __reverse reaction__ d. $3C_2H_4(g) \rightleftarrows C_6H_{12}(l)$

2. __$\Delta G = \Delta H - T\Delta S$__ a. Write an equation that shows the relationship between enthalpy, ΔH, entropy, ΔS, and free energy, ΔG.

   __negative__ b. For a reaction to occur spontaneously, the sign of ΔG should be ___.

3. Consider the following equation: $NH_3(g) + H_2O(l) \rightleftarrows NH_4^+(aq) + OH^-(aq) + $ energy

   __True__ a. The enthalpy factor favors the forward reaction. True or False?

   __reverse reaction__ b. The sign of $T\Delta S°$ is negative. This means the entropy factor favors the ___.

   c. Given that $\Delta G°$ for the above reaction in the forward direction is positive, which term is greater in magnitude and therefore predominates, $T\Delta S$ or ΔH?

   $T\Delta S > \Delta H$; ΔH is negative, but ΔG is positive, indicating the reaction is not

   spontaneous. The randomness of the reactants and the temperature of the reaction

   overshadow the exothermic factor.

4. Consider the following equation for the vaporization of water:

   $$H_2O(l) \rightleftarrows H_2O(g) \quad \Delta H = +40.65 \text{ kJ/mol at } 100°C$$

   __endothermic__ a. Is the forward reaction exothermic or endothermic?

   __reverse reaction__ b. Does the enthalpy factor favor the forward or reverse reaction?

   __forward reaction__ c. Does the entropy factor favor the forward or reverse reaction?

SECTION 2 continued

PROBLEMS Write the answer on the line to the left. Show all your work in the space provided.

5. Halogens can combine with other halogens to form several unstable compounds. Consider the following equation: $I_2(s) + Cl_2(g) \rightleftarrows 2ICl(g)$
ΔH_f^0 for the formation of ICl = +18.0 kJ/mol and $\Delta G^0 = -5.4$ kJ/mol.

<u>reverse reaction</u>   a. Is the forward or reverse reaction favored by the enthalpy factor?

<u>forward reaction</u>   b. Will the forward or reverse reaction occur spontaneously at standard conditions?

<u>forward reaction</u>   c. Is the forward or reverse reaction favored by the entropy factor?

<u>+23.4 kJ/(mol·K)</u>   d. Calculate the value of $T\Delta S$ for this system.

<u>0.0785 kJ/(mol·K)</u>   e. Calculate the value of ΔS for this system at 25°C.

6. Calculate the free-energy change for the reactions described by the equations below. Determine whether each reaction will be spontaneous or nonspontaneous.

<u>−51.0 kJ/mol; spontaneous</u>   a. $C(s) + 2H_2(g) \rightarrow CH_4(g)$
$\Delta S^0 = -80.7$ J/(mol·K), $\Delta H^0 = -75.0$ kJ/mol, $T = 298$ K

<u>195.8 kJ/mol; nonspontaneous</u>   b. $3Fe_2O_3(s) \rightarrow 2Fe_3O_4(s) + \frac{1}{2}O_2(g)$
$\Delta S^0 = 134.2$ J/(mol·K), $\Delta H^0 = 235.8$ kJ/mol, $T = 298$ K

Name _____ Date _____ Class _____

CHAPTER 16 REVIEW
Reaction Energy

MIXED REVIEW

SHORT ANSWER Answer the following questions in the space provided.

1. Describe Hess's law.

   The overall enthalpy change in a reaction is equal to the sum of enthalpy changes

   for the individual steps in the process.

2. What determines the amount of energy absorbed by a material when it is heated?

   Each material has its own unique specific heat value, which is the amount of energy

   it takes to raise the temperature of 1 g of the substance by 1°C, or 1 K. This value

   is dependent on the nature of the material, the mass of the sample, and the change

   in temperature

3. Describe what is meant by *enthalpy of combustion* and how a combustion calorimeter measures this enthalpy.

   The enthalpy of combustion is the enthalpy change that occurs during the complete

   combustion of 1 mol of substance and is measured using a combustion calorimeter.

   The sample is placed in the calorimeter and is ignited by an electric spark and

   burned in an atmosphere of pure oxygen. The energy generated by the combustion

   reaction warms the steel bomb and the water surrounding it. A thermometer

   measures the temperature change of the water and is used to calculated the energy

   that came from the reaction as heat.

MODERN CHEMISTRY   REACTION ENERGY **135**

Name _____ Date _____ Class _____

MIXED REVIEW continued

4. The following equation represents a reaction that is strongly favored in the forward direction:

$$2C_7H_5(NO_2)_3(l) + 12O_2(g) \rightarrow 14CO_2(g) + 5H_2O(g) + 3N_2O(g) + energy$$

a. Why would ΔG be negative in the above reaction?

Both energy and entropy factors favor the forward spontaneous reaction. The

reaction is exothermic, and there are more gas molecules in the products than in the

reactants.

PROBLEMS Write the answer on the line to the left. Show all your work in the space provided.

5. Consider the following equation and data: $2NO_2(g) \rightarrow N_2O_4(g)$

$$\Delta H_f^0 \text{ of } N_2O_4 = +9.2 \text{ kJ/mol}$$
$$\Delta H_f^0 \text{ of } NO_2 = +33.2 \text{ kJ/mol}$$
$$\Delta G^0 = -4.7 \text{ kJ/mol } N_2O_4$$

$\Delta H^0 = -57.2$ kJ/mol   Use Hess's law to calculate ΔH^0 for the above reaction.

6. _____2.26×10^4 J_____ Calculate the energy needed to raise the temperature of 180.0 g of water from 10.0°C to 40.0°C. The specific heat of water is 4.18 J/(K · g).

7. a. _____−233.5 kJ/mol_____ Calculate the change in Gibbs free energy for the following equation at 25°C.

$$2H_2O_2(l) \rightarrow 2H_2O(l) + O_2(g)$$

Given $\Delta H = -196.0$ kJ/mol
$\Delta S = +125.9$ J/mol

b. _____yes_____ Is this reaction spontaneous?

CHAPTER 17 REVIEW

Reaction Kinetics

SECTION 1

SHORT ANSWER Answer the following questions in the space provided.

1. Refer to the energy diagram below to answer the following questions.

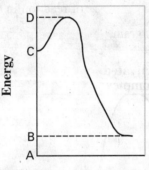

__d__ a. Which letter represents the energy of the activated complex?

(a) A (c) C
(b) B (d) D

__c__ b. Which letter represents the energy of the reactants?

(a) A (c) C
(b) B (d) D

__d__ c. Which of the following represents the quantity of activation energy for the forward reaction?

(a) the amount of energy at C minus the amount of energy at B
(b) the amount of energy at D minus the amount of energy at A
(c) the amount of energy at D minus the amount of energy at B
(d) the amount of energy at D minus the amount of energy at C

__c__ d. Which of the following represents the quantity of activation energy for the reverse reaction?

(a) the amount of energy at C minus the amount of energy at B
(b) the amount of energy at D minus the amount of energy at A
(c) the amount of energy at D minus the amount of energy at B
(d) the amount of energy at D minus the amount of energy at C

__b__ e. Which of the following represents the energy change for the forward reaction?

(a) the amount of energy at C minus the amount of energy at B
(b) the amount of energy at B minus the amount of energy at C
(c) the amount of energy at D minus the amount of energy at B
(d) the amount of energy at B minus the amount of energy at A

MODERN CHEMISTRY   REACTION KINETICS **137**

SECTION 1 continued

2. For the reaction described by the equation A + B → X, the activation energy for the forward direction equals 85 kJ/mol and the activation energy for the reverse direction equals 80 kJ/mol.

_____the product_____ **a.** Which has the greater energy content, the reactants or the product?

_____+5 kJ/mol_____ **b.** What is the enthalpy of reaction in the forward direction?

_____True_____ **c.** The enthalpy of reaction in the reverse direction is equal in magnitude but opposite in sign to the enthalpy of reaction in the forward direction. True or False?

3. Below is an incomplete energy diagram.

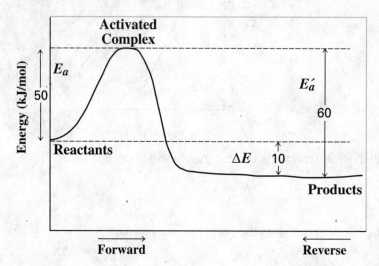

a. Use the following data to complete the diagram: $E_a = +50$ kJ/mol; $\Delta E_{forward} = -10$ kJ/mol. Label the reactants, products, ΔE, E_a, E_a', and the activated complex.

_____+60 kJ/mol_____ **b.** What is the value of E_a'?

4. It is proposed that ozone undergoes the following two-step mechanism in our upper atmosphere.

$$O_3(g) \rightarrow O_2(g) + O(g)$$
$$O_3(g) + O(g) \rightarrow 2O_2(g)$$

a. Identify any intermediates formed in the above equations.

Monatomic O is the intermediate formed.

b. Write the net equation.

$2O_3(g) \rightarrow 3O_2(g)$

_____exothermic_____ **c.** If ΔE is negative for the reaction in part **b**, what type of reaction is represented?

CHAPTER 17 REVIEW
Reaction Kinetics

SECTION 2

SHORT ANSWER Answer the following questions in the space provided.

1. Below is an energy diagram for a particular process. One curve represents the energy profile for the uncatalyzed reaction, and the other curve represents the energy profile for the catalyzed reaction.

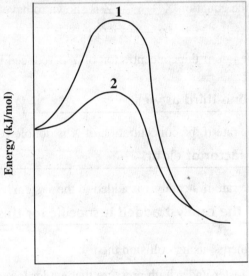

____a____ a. Which curve has the greater activation energy?

   (a) curve 1
   (b) curve 2
   (c) Both are equal.

____c____ b. Which curve has the greater energy change, ΔE?

   (a) curve 1
   (b) curve 2
   (c) Both are equal.

____b____ c. Which curve represents the catalyzed process?

   (a) curve 1
   (b) curve 2

   d. Explain your answer to part c.

   The catalyst forms an alternative activated complex that requires a lower activation

   energy, as represented by the lower curve.

SECTION 2 continued

2. Is it correct to say that a catalyst affects the speed of a reaction but does not take part in the reaction? Explain your answer.

   It is not correct. The catalyst does take part in the reaction. However, if it is used up in one step of the mechanism, it is regenerated in a later step. There is no net change in mass for the catalyst.

3. The reaction described by the equation X + Y → Z is shown to have the following rate law:

   $$R = k[X]^3[Y]$$

   a. What is the effect on the rate if the concentration of Y is reduced by one-third and [X] remains constant?
   The rate is reduced by one-third as well.

   b. What is the effect on the rate if the concentration of X is doubled and [Y] remains constant?
   The rate increases by a factor of eight.

   c. What is the effect on the rate if a catalyst is added to the system?
   The rate will increase if the catalyst added is specific for this reaction.

4. Explain the following statements, using collision theory:

   a. Gaseous reactants react faster under high pressure than under low pressure.
   At high pressure, gas molecules are more closely packed and collide more frequently. Thus, more-effective collisions occur per unit of time.

   b. Ionic compounds react faster when in solution than as solids.
   Ions in solution have more freedom of motion than do ions in a solid; therefore, they can collide with one another more frequently.

   c. A class of heterogeneous catalysts called surface catalysts work best as a fine powder.
   The fine powder has more surface area on which reactant particles can be absorbed and, in effect, increases the concentration of the reactants. An increase in concentration increases the number of effective collisions between reactant particles.

CHAPTER 17 REVIEW
Reaction Kinetics

MIXED REVIEW

SHORT ANSWER Answer the following questions in the space provided.

1. The reaction for the decomposition of hydrogen peroxide is $2H_2O_2(aq) \rightarrow 2H_2O(l) + O_2(g)$.

   List three ways to speed up the rate of decomposition. For each one, briefly explain why it is effective, based on collision theory.

   increase the concentration of hydrogen peroxide—allows more collisions per unit of

   time to occur

   increase the temperature of the solution—allows more energetic collisions per unit of

   time to occur

   stir the solution—exposes more reactant surface area, which allows more collisions

   per unit of time to occur

   add a catalyst—lowers the activation energy so that more-effective collisions can occur

2. An ingredient in smog is the gas NO. One reaction that controls the concentration of NO is

   $$H_2(g) + 2NO(g) \rightarrow H_2O(g) + N_2O(g).$$

   At high temperatures, doubling the concentration of H_2 doubles the rate of reaction, while doubling the concentration of NO increases the rate fourfold.

   Write a rate law for this reaction consistent with these data.

   $R = k[H_2][NO]^2$

3. Use the following chemical equation to answer the question below:

   $$Mg(s) + 2H_3O^+(aq) + Cl^-(aq) \rightarrow Mg^{2+}(aq) + 2Cl^-(aq) + H_2(g) + H_2O(l)$$

   If 0.048 g of magnesium completely reacts in 20 s, what is the average reaction rate in moles/second over that time interval?

   Average rate = 9.9×10^{-5} mol/s

MIXED REVIEW continued

PROBLEMS Write the answer on the line to the left. Show all your work in the space provided.

4. Answer the following questions using the energy diagram below.

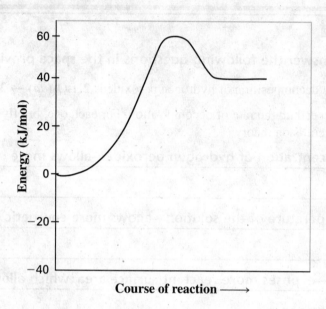

__endothermic__ a. Is the forward reaction represented by the curve exothermic or endothermic?

__+40 kJ/mol__ b. Estimate the magnitude and sign of $\Delta E_{forward}$.

__+20 kJ/mol__ c. Estimate E_a'.

A catalyst is added to the reaction, which lowers E_a by about 15 kJ/mol.

__speed up__ d. Does the forward reaction rate speed up or slow down?

__speed up__ e. Does the reverse reaction rate speed up or slow down?

__No__ f. Does $\Delta E_{forward}$ change from its value in part b?

5. a. Determine the overall balanced equation for a reaction having the following proposed mechanism:

Step 1: $2NO + H_2 \rightarrow N_2 + H_2O_2$ Slow
Step 2: $H_2 + H_2O_2 \rightarrow 2H_2O$ Fast

$2NO + 2H_2 \rightarrow N_2 + 2H_2O$

b. Which is the rate-determining step?
Step 1

c. What is the intermediate in the above reaction?
H_2O_2

CHAPTER 18 REVIEW
Chemical Equilibrium

SECTION 1

SHORT ANSWER Answer the following questions in the space provided.

1. Write the equilibrium expression for the following hypothetical equation:

$$3A(aq) + B(aq) \rightleftarrows 2C(aq) + 3D(aq)$$

$$K = \frac{[C]^2[D]^3}{[A]^3[B]}$$

2. a. Write the appropriate chemical equilibrium expression for each of the following equations. Include the value of K.

   (1) $N_2O_4(g) \rightleftarrows 2NO_2(g)$ $K = 0.1$

   $$\frac{[NO_2]^2}{[N_2O_4]} = 0.1$$

   (2) $NH_4OH(aq) \rightleftarrows NH_4^+(aq) + OH^-(aq)$ $K = 2 \times 10^{-5}$

   $$\frac{[NH_4^+][OH_2]}{[NH_4OH]} = 2 \times 10^{-5}$$

   (3) $H_2(g) + I_2(g) \rightleftarrows 2HI(g)$ $K = 54.0$

   $$\frac{[HI]^2}{[H_2][I_2]} = 54.0$$

   (4) $2SO_2(g) + O_2(g) \rightleftarrows 2SO_3(g)$ $K = 1.8 \times 10^{-2}$

   $$\frac{[SO_3]^2}{[SO_2]^2[O_2]} = 1.8 \times 10^{-2}$$

Name _____ Date _____ Class _____

SECTION 1 continued

_____system 3_____ **b.** Which of the four systems in part **a** proceeds furthest forward before equilibrium is established?

_____system 2_____ **c.** Which system contains mostly reactants at equilibrium?

3. a. Compare the rates of forward and reverse reactions when equilibrium has been reached.

The rate of the forward reaction equals the rate of the reverse reaction.

b. Describe what happens to the concentrations of reactants and products when chemical equilibrium has been reached.

The concentrations of the products and the reactants remain constant.

PROBLEMS Write the answer on the line to the left. Show all your work in the space provided.

4. _____1.1_____ Consider the following equation:

$$2NO(g) + O_2(g) \rightleftarrows 2NO_2(g)$$

At equilibrium, [NO] = 0.80 M, [O$_2$] = 0.50 M, and [NO$_2$] = 0.60 M. Calculate the value of K for this reaction.

5. _____0.0024_____ What is the K value for the following equation if the gaseous mixture in a 4 L container reaches equilibrium at 1000 K and contains 4.0 mol of N$_2$, 6.4 mol of H$_2$, and 0.40 mol of NH$_3$?

$$N_2(g) + 3H_2(g) \rightleftarrows 2NH_3(g)$$

CHAPTER 18 REVIEW
Chemical Equilibrium

SECTION 2

SHORT ANSWER Answer the following questions in the space provided.

1. __d__ Raising the temperature of any equilibrium system always
   - (a) favors the forward reaction.
   - (b) favors the reverse reaction.
   - (c) favors the exothermic reaction.
   - (d) favors the endothermic reaction.

2. Consider the following equilibrium equation: $CH_3OH(g) + 101 \text{ kJ} \rightleftarrows CO(g) + 2H_2(g)$.

   __b__ a. Increasing [CO] will
   - (a) increase $[H_2]$.
   - (b) decrease $[H_2]$.
   - (c) not change $[H_2]$.
   - (d) cause $[H_2]$ to fluctuate.

   __b__ b. Raising the temperature will cause the equilibrium of the system to
   - (a) favor the reverse reaction.
   - (b) favor the forward reaction.
   - (c) shift back and forth.
   - (d) remain as it was before.

   __a__ c. Raising the temperature will
   - (a) increase the value of K.
   - (b) decrease the value of K.
   - (c) not change the value of K.
   - (d) make the value of K fluctuate.

3. Consider the following equilibrium equation: $H_2O(g) + C(s) \rightleftarrows H_2(g) + CO(g) + \text{energy}$
   At equilibrium, which reaction will be favored (forward, reverse, or neither) when

   __reverse__ a. extra CO gas is introduced?

   __neither direction__ b. a catalyst is introduced?

   __forward__ c. the temperature of the system is lowered?

   __reverse__ d. the pressure on the system is increased due to a decrease in the container volume?

4. __c__ Silver chromate dissolves in water according to the following equation:

   $$Ag_2CrO_4(s) \rightleftarrows 2Ag^+(aq) + CrO_4^{2-}(aq)$$

   Which of these correctly represents the equilibrium expression for the above equation?

   (a) $\dfrac{2[Ag^+] + [CrO_4^{2-}]}{Ag_2CrO_4}$
   (b) $\dfrac{[Ag_2CrO_4]}{[Ag^+]^2[CrO_4^{2-}]}$
   (c) $\dfrac{[Ag^+]^2[CrO_4^{2-}]}{1}$
   (d) $\dfrac{[Ag^+]^2[CrO_4^{2-}]}{2[Ag_2CrO_4]}$

SECTION 2 continued

5. Are pure solids included in equilibrium expressions? Explain your answer.

   Pure solids are not included in equilibrium expressions because their concentrations do not change. Their constant value is incorporated into K.

6. A key step in manufacturing sulfuric acid is represented by the following equation:

   $$2SO_2(g) + O_2(g) \rightleftarrows 2SO_3(g) + 100 \text{ kJ/mol}$$

   To be economically viable, this process must yield as much SO_3 as possible in the shortest possible time. You are in charge of this manufacturing process.

   a. Would you impose a high pressure or a low pressure on the system? Explain your answer.

   Impose a high pressure so the reaction that produces fewer gas molecules will be favored, which is the forward reaction in this case.

   b. To maximize the yield of SO_3, should you keep the temperature high or low during the reaction?

   Keep the temperature low to favor the forward, exothermic reaction, but not so low as to slow the rate of the forward reaction too much.

   c. Will adding a catalyst change the yield of SO_3?

   No, a catalyst does not change the percentage yield.

7. The equation for an equilibrium system easily studied in a lab follows:

   $$2NO_2(g) \rightleftarrows N_2O_4(g)$$

   N_2O_4 gas is colorless, and NO_2 gas is dark brown. Lowering the temperature of the equilibrium mixture of gases reduces the intensity of the color.

   a. Is the forward or reverse reaction favored when the temperature is lowered?

   The forward reaction is favored. The color becomes less intense because the equilibrium shifts in the direction that produces the colorless gas, N_2O_4.

   b. Will the sign of ΔH be positive or negative if the temperature is lowered? Explain your answer.

   Negative; lowering the temperature favors the forward reaction, which is exothermic, so ΔH has a negative sign.

CHAPTER 18 REVIEW
Chemical Equilibrium

SECTION 3

SHORT ANSWER Answer the following questions in the space provided.

1. __a__ Lime juice turns litmus paper red, indicating that lime juice is
   (a) acidic.
   (b) basic.
   (c) neutral.
   (d) alkaline

2. __d__ Addition of the salt of a weak acid to a solution of the weak acid
   (a) lowers the concentration of the nonionized acid and the concentration of the H_3O^+ ion.
   (b) lowers the concentration of the nonionized acid and raises the concentration of the H_3O^+ ion.
   (c) raises the concentration of the nonionized acid and the concentration of the H_3O^+ ion.
   (d) raises the concentration of the nonionized acid and lowers the concentration of the H_3O^+ ion.

3. __b__ Salts of a weak acid and a strong base produce solutions that are
   (a) acidic only.
   (b) basic only.
   (c) neutral only.
   (d) either acidic, basic, or neutral.

4. __a__ If an acid is added to a solution of a weak base and its salt,
   (a) more water is formed and more weak base ionizes.
   (b) hydronium ion concentration decreases.
   (c) more hydroxide ion is formed.
   (d) more nonionized weak base is formed.

5. a. In the space below each of the following equations, correctly label the two conjugate acid-base pairs as *acid 1, acid 2, base 1,* and *base 2*.

   (a) $CO_3^{2-}(aq) + H_3O^+(aq) \rightleftarrows HCO_3^-(aq) + H_2O(l)$
       base 1   acid 2   acid 1   base 2

   (b) $HPO_4^{2-}(aq) + H_2O(l) \rightleftarrows OH^-(aq) + H_2PO_4^-(aq)$
       base 1   acid 2   base 2   acid 1

   __b__ b. Which reaction in part **a** is an example of hydrolysis?

   __a__ c. As the first reaction in part **a** proceeds, the pH of the solution
   (a) increases.   (c) stays at the same level.
   (b) decreases.   (d) fluctuates.

SECTION 3 continued

6. Write the formulas for the acid and the base that could form the salt $Ca(NO_3)_2$.

The acid is $HNO_3(aq)$ and the base is $Ca(OH)_2(aq)$.

7. Consider the following equation for the reaction of a weak base in water:

$$NH_3(aq) + H_2O(l) \rightleftharpoons NH_4^+(aq) + OH^-(aq)$$

Write the equilibrium expression for K.

$$K = \frac{[NH_4^+][OH^-]}{[NH_3]}$$

PROBLEMS Write the answer on the line to the left. Show all your work in the space provided.

8. An unknown acid X hydrolyzes according to the equation in part **a** below.

a. In the space below the equation, correctly label the two conjugate acid-base pairs in this system as *acid 1*, *acid 2*, *base 1*, and *base 2*.

$$HX(aq) + H_2O(l) \rightleftharpoons X^-(aq) + H_3O^+(aq)$$

acid 1   base 2   base 1   acid 2

b. Write the equilibrium expression for K_a for this system.

$$K_a = \frac{[X^-][H_3O^+]}{[HX]}$$

__1.0×10^{-8}__ **c.** Experiments show that at equilibrium $[H_3O^+] = [X^-] = 2.0 \times 10^{-5}$ mol/L and $[HX] = 4.0 \times 10^{-2}$ mol/L. Calculate the value of K_a based on these data.

CHAPTER 18 REVIEW
Chemical Equilibrium

SECTION 4

SHORT ANSWER Answer the following questions in the space provided.

1. Match the solution type on the right to the corresponding relationship between the ion product and the K_{sp} for that solution, listed on the left.

   __c__ The ion product exceeds the K_{sp}. (a) The solution is saturated; no more solid will dissolve.

   __a__ The ion product equals the K_{sp}. (b) The solution is unsaturated; no solid is present.

   __b__ The ion product is less than the K_{sp}. (c) The solution is supersaturated; solid may form if the solution is disturbed.

2. Silver carbonate, Ag_2CO_3, makes a saturated solution with $K_{sp} = 10^{-11}$.

   a. Write the equilibrium expression for the dissolution of Ag_2CO_3.

   $$K_{sp} = [Ag^+]^2[CO_3^{2-}]$$

   __reverse reaction__ b. In this system, will the foward or reverse reaction be favored if extra Ag^+ ions are added?

PROBLEMS Write the answer on the line to the left. Show all your work in the space provided.

3. When the ionic solid XCl_2 dissolves in pure water to make a saturated solution, experiments show that 2×10^{-3} mol/L of X^{2+} ions go into solution.

   __$XCl_2(s) \rightleftarrows X^{2+}(aq) + 2Cl^-(aq)$__ a. Write the equation showing the dissolution of XCl_2 and the corresponding equilibrium expression.

   __$K_{sp} = [X^{2+}][Cl^-]^2$__

   __3×10^{-8}__ b. Calculate the value of K_{sp} for XCl_2.

SECTION 4 continued

___less soluble___   **c.** Refer to **Table 3** on page 615 of the text. Would XCl_2 be more soluble or less soluble than $PbCl_2$ at the same temperature?

4. The solubility of Ag_3PO_4 is 2.1×10^{-4} g/100. g.

$\underline{Ag_3PO_4(s) \rightleftarrows 3Ag^+(aq) + PO_4^{3-}(aq)}$ **a.** Write the equation showing the dissolution of this ionic solid.

___5.0×10^{-6} M___   **b.** Calculate the molarity of this saturated solution.

___1.7×10^{-20}___   **c.** What is the value of K_{sp} for this system?

5. As $PbCl_2$ dissolves, $[Pb^{2+}] = 2.0 \times 10^{-1}$ mol/L and $[Cl^-] = 1.5 \times 10^{-2}$ mol/L.

___$K_{sp} = [Pb^{2+}][Cl^-]^2$___   **a.** Write the equilibrium expression for the dissolution of $PbCl_2$.

___4.5×10^{-5}___   **b.** Compute the ion product, using the data given above.

CHAPTER 18 REVIEW
Chemical Equilibrium

MIXED REVIEW

SHORT ANSWER Answer the following questions in the space provided.

1. Consider the following equilibrium equation:

$$N_2(g) + 2O_2(g) \rightleftarrows 2NO_2(g); \Delta H = +33 \text{ kJ/mol}$$

   At equilibrium, which reaction is favored when

   __reverse reaction__ a. some N_2 is removed?

   __neither reaction__ b. a catalyst is introduced?

   __forward reaction__ c. pressure on the system increases due to a decrease in the volume?

   __forward reaction__ d. the temperature of the system is increased?

2. Ammonia gas dissolves in water according to the following equation:

$$NH_3(g) + H_2O(l) \rightleftarrows NH_4^+(aq) + OH^-(aq) + \text{energy}; K = 1.8 \times 10^{-5}$$

   __base__ a. Is aqueous ammonia an acid or a base?

   __Yes__ b. Is the equation given above an example of hydrolysis?

   __reverse reaction__ c. For the given value of K, does the equilibrium favor the forward or reverse reaction?

PROBLEMS Write the answer on the line to the left. Show all your work in the space provided.

3. Formic acid, HCOOH, is a weak acid present in the venom of red-ants. At equilibrium, [HCOOH] = 2.00 M, [HCOO$^-$] = 4.0 × 10^{-1} M, and [H$_3$O$^+$] = 9.0 × 10^{-4} M.

   $K_a = \dfrac{[H_3O^+][HCOO^-]}{[HCOOH]}$ a. Write the equilibrium expression for the ionization of formic acid.

   __1.8×10^{-4}__ b. Calculate the value of K_a for this acid.

MIXED REVIEW continued

4. HF hydrolyzes according to the following equation:
$$HF(aq) + H_2O(l) \rightleftharpoons H_3O^+(aq) + F^-(aq)$$

When 0.0300 mol of HF dissolves in 1.00 L of water, the solution quickly ionizes to reach equilibrium. At equilibrium, the remaining [HF] = 0.0270 M.

____0.0030 mol/L____ a. How many moles of HF ionize per liter of water to reach equilibrium?

____0.0030 mol/L for both____ b. What are $[F^-]$ and $[H_3O^+]$?

____3.3×10^{-4}____ c. What is the value of K_a for HF?

5. Refer to **Table 3** on page 615 of the text. $CaSO_4(s)$ is only slightly soluble in water.

____$CaSO_4(s) \rightleftharpoons Ca^{2+}(aq) + SO_4^{2-}(aq)$____
____$K_{sp} = [Ca^{2+}][SO_4^{2-}] = 9.1 \times 10^{-6}$____ a. Write the equilibrium equation and equilibrium expression for the dissolution of $CaSO_4(s)$ with the K_{sp} value.

____4.1×10^{-2} g/100. g H_2O____ b. Determine the solubility of $CaSO_4$ at 25°C in grams per 100. g H_2O.

CHAPTER 19 REVIEW
Oxidation-Reduction Reactions

SECTION 1

SHORT ANSWER Answer the following questions in the space provided.

1. __a__ All the following equations involve redox reactions *except*
   (a) $CaO + H_2O \rightarrow Ca(OH)_2$.
   (b) $2SO_2 + O_2 \rightarrow 2SO_3$.
   (c) $2HgO \rightarrow 2Hg + O_2$.
   (d) $SnCl_4 + 2FeCl_2 \rightarrow 2FeCl_3 + SnCl_2$.

2. Assign the correct oxidation number to the individual atom or ion below.

   __+4__ a. Mn in MnO_2
   __0__ b. S in S_8
   __−1__ c. Cl in $CaCl_2$
   __+5__ d. I in IO_3^-
   __+4__ e. C in HCO_3^-
   __+3__ f. Fe in $Fe_2(SO_4)_3$
   __+6__ g. S in $Fe_2(SO_4)_3$

3. In each of the following half-reactions, determine the value of x.

   __8__ a. $S^{6+} + x\,e^- \rightarrow S^{2-}$
   __1−__ b. $2Br^x \rightarrow Br_2 + 2e^-$
   __2+__ c. $Sn^{4+} + 2e^- \rightarrow Sn^x$
   __a, c__ d. Which of the above half-reactions represent reduction processes?

4. Give examples, other than those listed in **Table 1** on page 631 of the text, for the following:
   Answers may vary.

   __NaH or CaH$_2$__ a. a compound containing H in a −1 oxidation state
   __Na$_2$O$_2$ or BaO$_2$__ b. a peroxide
   __SO$_3^{2-}$ or HSO$_3^-$__ c. a polyatomic ion in which the oxidation number for S is +4
   __F$_2$__ d. a substance in which the oxidation number for F is not −1

MODERN CHEMISTRY

SECTION 1 continued

5. OILRIG is a mnemonic device often used by students to help them understand redox reactions.

"*Oxidation is loss, reduction is gain.*"

Explain what that phrase means—loss and gain of what?

Oxidation involves losing (or ejecting) electrons; reduction involves gaining electrons.

6. For each of the reactions described by the following equations, state whether or not any oxidation and reduction is occurring, and write the oxidation-reduction half-reactions for those cases in which redox does occur.

   a. $Ca(OH)_2(aq) + 2HCl(aq) \rightarrow CaCl_2(aq) + 2H_2O(l)$

   no redox occurring

   b. $CH_4(g) + 2O_2(g) \rightarrow CO_2(g) + 2H_2O(g)$

   yes; oxidation: $C^{4-} \rightarrow C^{4+} + 8e^-$; reduction: $O + 2e^- \rightarrow O^{2-}$

   c. $2Al(s) + 3CuCl_2(aq) \rightarrow 2AlCl_3(aq) + 3Cu(s)$

   yes; oxidation: $Al \rightarrow Al^{3+} + 3e^-$; reduction: $Cu^{2+} + 2e^- \rightarrow Cu$

7. I^- is converted into I_2 by the addition of an aqueous solution of $KMnO_4$ to an aqueous solution of KI.

   a. What is the oxidation number assigned to I_2?

   0

   b. The conversion of I^- to I_2 is a(n) _____ reaction.

   oxidation

   c. How many electrons are lost when 1 mol of I_2 is formed from I^-?

   2 mol of electrons

CHAPTER 19 REVIEW
Oxidation-Reduction Reactions

SECTION 2

SHORT ANSWER Answer the following questions in the space provided.

1. __c__ All of the following should be done in the process of balancing redox equations *except*

    (a) adjusting coefficients to balance atoms.
    (b) adjusting coefficients in electron equations to balance numbers of electrons lost and gained.
    (c) adjusting subscripts to balance atoms.
    (d) writing two separate electron equations.

2. MnO_4^- can be reduced to MnO_2.

    __MnO_4^- : +7, MnO_2 : +4__   a. Assign the oxidation number to Mn in these two species.

    __$3e^-$__   b. How many electrons are gained per Mn atom in this reduction?

    __$9.0 \times 10^{23}\ e^-$__   c. If 0.50 mol of MnO_4^- is reduced, how many electrons are gained?

3. Iodide ions can be oxidized to form iodine. Write the balanced oxidation half-reaction for the oxidation of iodide to iodine.

    $2I^- \rightarrow I_2 + 2e^-$

4. Some bleaches contain aqueous chlorine as the active ingredient. Aqueous chlorine is made by dissolving chlorine gas in water. Aqueous chlorine is capable of oxidizing iron(II) ions to iron(III) ions. When iron(II) ions are oxidized, chloride ions are formed.

    a. Write equations for the two half-reactions involved. Label them *oxidation* or *reduction*.

    oxidation : $Fe^{2+} \rightarrow Fe^{3+} + 1e^-$

    reduction : $Cl_2 + 2e^- \rightarrow 2Cl^-$

    b. Write the balanced ionic equation for the redox reaction between aqueous chlorine and iron(II).

    $Cl_2(aq) + 2Fe^{2+}(aq) \rightarrow 2Fe^{3+}(aq) + 2Cl^-(aq)$

    c. Show that the equation in part **b** is balanced by charge.

    total charge on the left = 0 + (+4) = +4; total charge on the right = +6 + (−2) = +4

SECTION 2 continued

5. Write the equations for the oxidation and reduction half-reactions for the redox reactions below, and then balance the reaction equations.

   a. $MnO_2(s) + HCl(aq) \rightarrow MnCl_2(aq) + Cl_2(g) + H_2O(l)$

   oxidation: $2Cl^- \rightarrow Cl_2 + 2e^-$
   reduction: $MnO_2 + 4H^+ + 2e^- \rightarrow Mn^{2+} + 2H_2O$
   $MnO_2(s) + 4HCl(aq) \rightarrow MnCl_2(aq) + Cl_2(g) + 2H_2O(l)$

   b. $S(s) + HNO_3(aq) \rightarrow SO_3(g) + H_2O(l) + NO_2(g)$

   oxidation: $(S + 3H_2O \rightarrow SO_3 + 6H^+ + 6e^-) \times 1$
   reduction: $(1e^- + NO_3^- + 2H^+ \rightarrow NO_2 + H_2O) \times 6$
   $S(s) + 6HNO_3(aq) \rightarrow SO_3(g) + 3H_2O(l) + 6NO_2(g)$

   c. $H_2C_2O_4(aq) + K_2CrO_4(aq) + HCl(aq) \rightarrow CrCl_3(aq) + KCl(aq) + H_2O(l) + CO_2(g)$

   oxidation: $(C_2O_4^{2-} \rightarrow 2CO_2 + 2e^-) \times 3$
   reduction: $(CrO_4^{2-} + 8H^+ + 3e^- \rightarrow Cr^{3+} + 4H_2O) \times 2$
   $3H_2C_2O_4(aq) + 2K_2CrO_4(aq) + 10HCl(aq) \rightarrow$
   $2CrCl_3(aq) + 4KCl(aq) + 8H_2O(l) + 6CO_2(g)$

CHAPTER 19 REVIEW
Oxidation-Reduction Reactions

SECTION 3

SHORT ANSWER Answer the following questions in the space provided.

1. For each of the following, identify the stronger oxidizing or reducing agent. (Refer to **Table 3** on page 643 of the text.)

    __Ca__   a. Ca or Cu as a reducing agent

    __Ag^+__   b. Ag^+ or Na^+ as an oxidizing agent

    __Fe^{3+}__   c. Fe^{3+} or Fe^{2+} as an oxidizing agent

2. For each of the following incomplete equations, state whether a redox reaction is likely to occur. (Refer to **Table 3** on page 643 of the text.)

    __Yes__   a. $Mg + Sn^{2+} \rightarrow$

    __No__   b. $Ag + Cu^{2+} \rightarrow$

    __Yes__   c. $Br_2 + I^- \rightarrow$

3. Label each of the following statements about redox as True or False.

    __True__   a. A strong oxidizing agent is itself readily reduced.

    __True__   b. In disproportionation, one chemical acts as both an oxidizing agent and a reducing agent in the same process.

    __False__   c. The number of moles of chemical oxidized must equal the number of moles of chemical reduced.

4. Solutions of Fe^{2+} are fairly unstable, in part because they can undergo disproportionation, as shown by the following unbalanced equation:

    $$Fe^{2+} \rightarrow Fe^{3+} + Fe$$

    a. Balance the above equation.

    __$3Fe^{2+} \rightarrow 2Fe^{3+} + Fe$__

    __0.072 mol__   b. If the reaction described above produces 0.036 mol of Fe, how many moles of Fe^{3+} will form?

SECTION 3 continued

5. Oxygen gas is a powerful oxidizing agent.

_____0_____ **a.** Assign the oxidation number to O_2.

_____−2_____ **b.** What does oxygen's oxidation number usually become when it functions as an oxidizing agent?

c. Approximately where would you place O_2 in the list of oxidizing agents in **Table 3** on page 643 of the text?

at the bottom right, near F_2, but above it

d. Describe the changes in oxidation states that occur in carbon and oxygen, and identify the oxidizing and reducing agents, in the combustion reaction described by the following equation:

$$C_6H_{12}O_6(s) + 6O_2(g) \rightarrow 6CO_2(g) + 6H_2O(l)$$

Each O atom in O_2 changes from the oxidation state of 0 to −2 in CO_2 and H_2O.

Oxygen is reduced; therefore, it is the oxidizing agent. Each C atom in $C_6H_{12}O_6$ as a

reactant is in the oxidation state of 0, and in CO_2 as a product, it is +4. It is

oxidized and is therefore the reducing agent.

6. An example of disproportionation is the slow decomposition of aqueous chlorine, $Cl_2(aq)$, represented by the following unbalanced equation:

$$Cl_2(aq) + H_2O(l) \rightarrow ClO^-(aq) + Cl^-(aq) + H^+(aq)$$

a. Show that the oxygen and hydrogen atoms in the above reaction are not changing oxidation states.

O has an oxidation number of −2 when it is part of both H_2O and ClO^-.

H has an oxidation number of +1 when it is part of H_2O and also as a H^+ ion.

b. Show the changes in the oxidation states of chlorine as this reaction proceeds.

$Cl^0 \rightarrow Cl^+ + 1e^-$ in oxidation to ClO^-; $Cl^0 + 1e^- \rightarrow Cl^-$ in reduction to Cl^-

_____$1e^-$_____ **c.** In the oxidation reaction, how many electrons are transferred per Cl atom?

_____$1e^-$_____ **d.** In the reduction reaction, how many electrons are transferred per Cl atom?

e. What must be the ratio of ClO^- to Cl^- in the above reaction? Explain your answer.

1 ClO^- to 1 Cl^- to make the number of electrons given off and gained equal

f. Balance the equation for the decomposition of $Cl_2(aq)$.

$Cl_2(aq) + H_2O \rightarrow ClO^-(aq) + Cl^-(aq) + 2H^+(aq)$

Name _____ Date _____ Class _____

CHAPTER 19 REVIEW
Oxidation-Reduction Reactions

MIXED REVIEW

SHORT ANSWER Answer the following questions in the space provided.

1. Label the following descriptions of reactions *oxidation*, *reduction*, or *disproportionation*.

   __disproportionation__ a. conversion of Na_2O_2 to NaO and O_2

   __oxidation__ b. conversion of Br^- to Br_2

   __oxidation__ c. conversion of Fe^{2+} to Fe^{3+}

   __reduction__ d. a change in the oxidation number in a negative direction

PROBLEMS Write the answer on the line to the left. Show all your work in the space provided.

2. Consider the following unbalanced equation:

   $KMnO_4(aq) + HCl(aq) + Al(s) \rightarrow AlCl_3(aq) + MnCl_2(aq) + KCl(aq) + H_2O(l)$

   a. Write the oxidation and reduction half-reactions. Label each half-reaction *oxidation* or *reduction*.

   oxidation: $Al \rightarrow Al^{3+} + 3e^-$

   reduction: $MnO_4^- + 8H^+ + 5e^- \rightarrow Mn^{2+} + 4H_2O$

   b. Balance the equation using the seven-step procedure shown on pages 637–639 of the text.

   $(MnO_4^- + 8H^+ + 5e^- \rightarrow Mn^{2+} + 4H_2O) \times 3$
   $(Al \rightarrow Al^{3+} + 3e^-) \times 5$
   $3KMnO_4 + 24HCl + 5Al \rightarrow 5AlCl_3 + 3MnCl_2 + 3KCl + 12H_2O$

   __MnO_4^-__ c. Identify the oxidizing agent in this system.

MODERN CHEMISTRY

MIXED REVIEW continued

3. Consider the unbalanced ionic equation $ClO^- + H^+ \rightarrow Cl_2 + ClO_3^- + H_2O$.

   a. Assign the oxidation number to each element.

   In ClO^-, the oxidation number for Cl is +1 and for O is −2; in H^+, the oxidation number for H is +1. In Cl_2, the oxidation number for Cl is 0; in ClO_3^-, the oxidation number for Cl is +5, and for O is −2; in H_2O, the oxidation number for H is +1, and for O is −2.

   _____ 4 e^- _____ b. How many electrons are given up by each Cl atom as it is oxidized?

   _____ 1 e^- _____ c. How many electrons are gained by each Cl atom as it is reduced?

   _____ Yes _____ d. Is this an example of disproportionation?

   e. Balance the above equation, using the method of your choice.

   $5ClO^- + 4H^+ \rightarrow 2Cl_2 + ClO_3^- + 2H_2O$

CHAPTER 20 REVIEW
Electrochemistry

SECTION 1

SHORT ANSWER Answer the following questions in the space provided.

1. The following reaction takes place in an electrochemical cell:

$$Cu^{2+}(aq) + Zn(s) \rightarrow Cu(s) + Zn^{2+}(aq)$$

a. Which electrode is the anode?

the zinc electrode

b. Which electrode is the cathode?

the copper electrode

c. Write the cell notation for this system.

$Zn(s) \mid Zn^{2+}(aq) \parallel Cu^{2+}(aq) \mid Cu(s)$

d. Write equations for the half-reactions that occur at each electrode and label each reaction.

at the anode, $Zn(s) \rightarrow Zn^{2+}(aq) + 2e^-$; at the cathode, $Cu^{2+}(aq) + 2e^- \rightarrow Cu(s)$

2. __b__ Energy will be released in the form of heat when

(a) reactants in a spontaneous energy-releasing redox reaction are connected by a wire.
(b) reactants in a spontaneous energy-releasing redox reaction are in direct contact.
(c) copper atoms are deposited on an anode.
(d) electrochemical half-cells are isolated from one another.

3. __d__ An electrochemical cell is constructed using the reaction of chromium metal and iron(II) ion, as follows:

$$2Cr(s) + 3Fe^{2+}(aq) \rightarrow 2Cr^{3+} + 3Fe(s)$$

Which statement best describes this system?

(a) Electrons flow from the iron electrode to the chromium electrode.
(b) Energy is released.
(c) Negative ions move through the salt bridge from the chromium half-cell to the iron half-cell.
(d) Negative ions move through the salt bridge from the iron half-cell to the chromium half-cell.

SECTION 1 continued

4. Below is a diagram of an electrochemical cell.

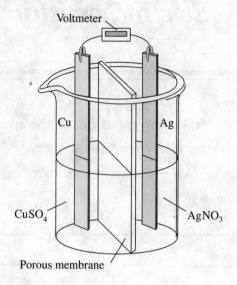

a. Write the anode half-reaction.

$Cu(s) \rightarrow Cu^{2+}(aq) + 2e^-$

b. Write the cathode half-reaction.

$Ag^+(aq) + 1e^- \rightarrow Ag(s)$

c. Write the balanced overall cell reaction.

$Cu(s) + 2Ag^+(aq) \rightarrow Cu^{2+}(aq) + 2Ag(s)$

__clockwise__ **d.** Do electrons within the electrochemical cell travel through the voltmeter in a clockwise or a counterclockwise direction, as represented in the diagram?

__right to left__ **e.** In what direction do anions pass through the porous membrane, as represented in the diagram?

CHAPTER 20 REVIEW
Electrochemistry

SECTION 2

SHORT ANSWER Answer the following questions in the space provided.

1. __c__ In a voltaic cell, transfer of charge through the external wires occurs by means of
   - (a) ionization.
   - (b) ion movement.
   - (c) electron movement.
   - (d) proton movement.

2. __c__ All the following claims about voltaic cells are true *except*
   - (a) $E°_{cell}$ is positive.
   - (b) The redox reaction in the cell occurs without the addition of electric energy.
   - (c) Electrical energy is converted to chemical energy.
   - (d) Chemical energy is converted to electrical energy.

3. Use **Table 1** on page 664 of the text to find $E°$ for the following:

   __+0.56 V__ a. the reduction of MnO_4^{1-} to MnO_4^{2-}

   __+0.74 V__ b. the oxidation of Cr to Cr^{3+}

   __0__ c. the reaction within the SHE

   __+0.29 V__ d. $Cl_2 + 2Br^- \rightarrow 2Cl^- + Br_2$

4. Why does a zinc coating protect steel from corrosion?
   Zinc is more easily oxidized than iron and reacts before the iron is oxidized.

5. Which two types of batteries share the following anode half-reaction?

   $$Zn(s) + 2OH^-(aq) \rightarrow Zn(OH)_2(s) + 2e^-$$

   alkaline dry cell and mercury

6. Complete the following sentences:

   Corrosion acts as a voltaic cell because oxidation and reduction reactions occur __spontaneously__ at different sites. The two half-cells are connected by a __metal conductor__, which allows electrons to flow.

Name _____ Date _____ Class _____

SECTION 2 continued

PROBLEM   Write the answer on the line to the left. Show all your work in the space provided.

7. Use **Table 1** on p. 664 of the text to find E^0 for the following voltaic cells:

   __+1.26 V__   a. $Al(s) \mid Al^{3+}(aq) \parallel Cd^{2+}(aq) \mid Cd(s)$

   __+0.28 V__   b. $Fe(s) \mid Fe^{2+}(aq) \parallel Pb^{2+}(aq) \mid Pb(s)$

   __+0.88 V__   c. $6I^-(aq) + 2Au^{3+}(aq) \rightarrow 3I_2(s) + 2Au(s)$

8. A voltmeter connected to a copper-silver voltaic cell reads +0.46 V. The silver is then replaced with metal X and its 2+ ion. A new voltage reading shows that the direction of the current has reversed, and the voltmeter reads +0.74 V. Use data from **Table 1** on page 664 of the text to answer.

   __−0.40 V__   a. Calculate the reduction potential of metal X.

   __Cd__   b. Predict the identity of metal X.

CHAPTER 20 REVIEW
Electrochemistry

SECTION 3

SHORT ANSWER Answer the following questions in the space provided.

1. Label each of the following statements as applying to a *voltaic cell*, an *electrolytic cell*, or *both*:

   both a. The cell reaction involves oxidation and reduction.

   voltaic cell b. The cell reaction proceeds spontaneously.

   electrolytic cell c. The cell reaction is endothermic.

   voltaic cell d. The cell reaction converts chemical energy into electrical energy.

   electrolytic cell e. The cell reaction converts electrical energy into chemical energy.

   both f. The cell contains both a cathode and an anode.

2. **a** In an electrolytic cell, oxidation takes place

   (a) at the anode.
   (b) at the cathode.
   (c) via the salt bridge.
   (d) at the positive electrode.

3. **c** Which atom forms an ion that would always migrate toward the cathode in an electrolytic cell?

   (a) F
   (b) I
   (c) Cu
   (d) Cl

4. An electrolytic process in which solid metal is deposited on a surface is called **electroplating**.

5. When a rechargeable camera battery is being recharged, the cell acts as a(n) **electrolytic** cell and converts **electrical** energy into **chemical** energy. When the battery is being used to power the camera, it acts as a(n) **voltaic** cell and converts **chemical** energy into **electrical** energy.

SECTION 3 continued

6. Using **Figure 14** on p. 668 as a guide, describe how you would electroplate gold, Au, onto a metal object from a solution of $Au(CN)_3$. Include in your discussion, the equation for the reaction that plates the gold.

   You will need a power source, wire leads with clips, a strip of gold, and a beaker.

   Pour the $Au(CN)_3$ solution into the beaker, make the object you want to plate the

   cathode and the gold strip the anode. The reaction that plates the gold is

   $Au^{3+}(aq) + 3e^- \rightarrow Au(s)$.

7. Explain why aluminum recycling is less expensive than the extraction of aluminum metal from bauxite ore.

   The extraction of aluminum from bauxite ore involves an electrolytic process that

   requires a large input of electrical energy. Aluminum can be recycled at less than

   one-tenth the cost of this process.

8. Label the following statements about the electrolysis of water as True or False.

False	The process is spontaneous.
True	Hydrogen gas is formed at the cathode.
True	Oxygen gas is formed at the anode.
False	Electric energy is generated.

CHAPTER 20 REVIEW

Electrochemistry

MIXED REVIEW

SHORT ANSWER Answer the following questions in the space provided.

1. __d__ An electrochemical cell consists of two electrodes separated by a(n)
   - (a) anode.
   - (b) cathode.
   - (c) voltage.
   - (d) electrolyte.

2. __d__ When a car battery is charging,
   - (a) electrical energy is converted into energy of motion.
   - (b) energy of motion is converted into electrical energy.
   - (c) chemical energy is converted into electrical energy.
   - (d) electrical energy is converted into chemical energy.

3. __a__ Electroplating is an application of
   - (a) electrolytic cell reactions.
   - (b) fuel cell reactions.
   - (c) auto-oxidation reactions.
   - (d) galvanic reactions.

4. __b__ A major benefit of electroplating is that it
   - (a) increases concentrations of toxic wastes.
   - (b) protects metals from corrosion.
   - (c) saves time.
   - (d) leads to a buildup of impurities.

5. __b__ The transfer of charge through the electrolyte solution occurs by means of
   - (a) ionization.
   - (b) ion movement.
   - (c) electron movement.
   - (d) proton movement.

Name _____ Date _____ Class _____

MIXED REVIEW continued

6. A voltaic cell is constructed that reacts according to the following equation:

$$Mg(s) + 2H^+(aq) \rightarrow Mg^{2+} + H_2(g)$$

   a. Write equations for the half-reactions that occur in this cell.

   $Mg(s) \rightarrow Mg^{2+}(aq) + 2e^-$

   $2H^+(aq) + 2e^- \rightarrow H_2(g)$

   b. Which half-reaction occurs in the anode half-cell?

   $Mg(s) \rightarrow Mg^{2+}(aq) + 2e^-$

   c. Write the cell notation for this cell.

   $Mg(s) \mid Mg^{2+}(aq) \parallel H^+(aq) \mid H_2(g)$

   d. Electrons flow through the wire from the __magnesium__ electrode to the __hydrogen__ electrode. Positive ions move from the __magnesium__ half-cell to the __hydrogen__ half-cell.

PROBLEM Write the answer on the line to the left. Show all your work in the space provided.

_____+2.14V_____ 7. Refer to **Table 1** on page 664 of the text. What is the voltage of the cell for the following reaction?

$$Mg + Ni(NO_3)_2 \rightarrow Ni + Mg(NO_3)_2$$

CHAPTER 21 REVIEW
Nuclear Chemistry

SECTION 1

SHORT ANSWER Answer the following questions in the space provided.

1. __b__ Based on the information about the three elementary particles on page 683 of the text, which has the greatest mass?
   - (a) the proton
   - (b) the neutron
   - (c) the electron
   - (d) They all have the same mass.

2. __a__ The force that keeps nucleons together is
   - (a) a strong nuclear force.
   - (b) a weak nuclear force.
   - (c) an electromagnetic force.
   - (d) a gravitational force.

3. __d__ The stability of a nucleus is most affected by the
   - (a) number of neutrons.
   - (b) number of protons.
   - (c) number of electrons.
   - (d) ratio of neutrons to protons.

4. __b__ If an atom should form from its constituent particles,
   - (a) matter is lost and energy is taken in.
   - (b) matter is lost and energy is released.
   - (c) matter is gained and energy is taken in.
   - (d) matter is gained and energy is released.

5. __b__ For atoms of a given mass number, those with greater mass defects, have
   - (a) smaller binding energies per nucleon.
   - (b) greater binding energies per nucleon.
   - (c) the same binding energies per nucleon as those with smaller mass defects.
   - (d) variable binding energies per nucleon.

6. Based on **Figure 1** on page 684 of the text, which isotope of He, helium-3 or helium-4,
   __helium-3__ a. has the smaller binding energy per nucleon?
   __helium-4__ b. is more stable to nuclear changes?

7. The number of neutrons in an atom of magnesium-25 is __13__.

8. Nuclides of the same element have the same __atomic number__.

MODERN CHEMISTRY   NUCLEAR CHEMISTRY **169**

Name _____ Date _____ Class _____

SECTION 1 continued

9. Atom X has 50 nucleons and a binding energy of 4.2×10^{-11} J. Atom Z has 80 nucleons and a binding energy of 8.4×10^{-11} J.

_____**True**_____ **a.** The mass defect of Z is twice that of X. True or False?

_____**atom Z**_____ **b.** Which atom has the greater binding energy per nucleon?

_____**atom Z**_____ **c.** Which atom is likely to be more stable to nuclear transmutations?

10. Identify the missing term in each of the following nuclear equations. Write the element's symbol, its atomic number, and its mass number.

_____$^{14}_{7}N$_____ **a.** $^{14}_{6}C \rightarrow {}^{0}_{-1}e + \underline{}$

_____$^{60}_{28}Ni$_____ **b.** $^{63}_{29}Cu + {}^{1}_{1}H \rightarrow \underline{} + {}^{4}_{2}He$

11. Write the equation that shows the equivalency of mass and energy.

$E = mc^2$

12. Consider the two nuclides $^{56}_{26}Fe$ and $^{14}_{6}C$.

a. Determine the number of protons in each nucleus.

Iron-56 has 26 protons; carbon-14 has 6 protons.

b. Determine the number of neutrons in each nucleus.

Iron-56 has 30 neutrons; carbon-14 has 8 neutrons.

c. Determine whether the $^{56}_{26}Fe$ nuclide is likely to be stable or unstable, based on its position in the band of stability shown in **Figure 2** on page 685 of the text.

Iron-56 is likely to be stable.

PROBLEM Write the answer on the line to the left. Show all your work in the space provided.

13. ___**0.172 46 amu**___ Neon-20 is a stable isotope of neon. Its actual mass has been found to be 19.992 44 amu. Determine the mass defect in this nuclide.

CHAPTER 21 REVIEW
Nuclear Chemistry

SECTION 2

SHORT ANSWER Answer the following questions in the space provided.

1. __a__ The nuclear equation $^{210}_{84}Po \rightarrow \,^{206}_{82}Pb + \,^{4}_{2}He$ is an example of an equation that represents
   (a) alpha emission.
   (b) beta emission.
   (c) positron emission.
   (d) electron capture.

2. __d__ When $^{a}_{b}Z$ undergoes electron capture to form a new element X, which of the following best represents the product formed?
   (a) $^{a-1}_{b}X$
   (b) $^{a+1}_{b}X$
   (c) $^{a}_{b+1}X$
   (d) $^{a}_{b-1}X$

3. __a__ Which of the following best represents the fraction of a radioactive sample that remains after four half-lives have occurred?
   (a) $\left(\frac{1}{2}\right)^4$
   (c) $\left(\frac{1}{4}\right)$
   (b) $\left(\frac{1}{2}\right) \times 4$
   (d) $\left(\frac{1}{4}\right)^2 \times 4$

4. Match the nuclear symbol on the right to the name of the corresponding particle on the left.

   __c__ beta particle   (a) $^{1}_{1}p$
   __a__ proton   (b) $^{4}_{2}He$
   __d__ positron   (c) $^{0}_{-1}\beta$
   __b__ alpha particle   (d) $^{0}_{+1}\beta$

5. Label each of the following statements as True or False. Consider the U-238 decay series on page 692 of the text. For the series of decays from U-234 to Po-218, each nuclide

   __False__ a. shares the same atomic number
   __True__   b. differs in mass number from others by multiples of 4
   __True__   c. has a unique atomic number
   __False__ d. differs in atomic number from others by multiples of 4

Name _____ Date _____ Class _____

SECTION 2 continued

6. _____$^{235}_{92}U$_____ Identify the missing term in the following nuclear equation. Write the element's symbol, its atomic number, and its mass number.

$$\underline{\,?\,} \rightarrow {}^{231}_{90}Th + {}^{4}_{2}He$$

7. Lead-210 undergoes beta emission. Write the nuclear equation showing this transmutation.

$${}^{210}_{82}Pb \rightarrow {}^{0}_{-1}\beta + {}^{210}_{83}Bi$$

8. Einsteinium is a transuranium element. Einsteinium-247 can be prepared by bombarding uranium-238 with nitrogen-14 nuclei, releasing several neutrons, as shown by the following equation:

$$ {}^{238}_{92}U + {}^{14}_{7}N \rightarrow {}^{247}_{99}Es + x\,{}^{1}_{0}n$$

What must be the value of x in the above equation? Explain your reasoning.

The value of x is 5. The total mass of the reactants is 252 amu, therefore the total mass of the products must be 252. Then 252 − 247 leaves 5 amu for neutrons at 1 amu each.

PROBLEMS Write the answer on the line to the left. Show all your work in the space provided.

9. ___42.9 days___ Phosphorus-32 has a half-life of 14.3 days. How many days will it take for a sample of phosphorus-32 to decay to one-eighth its original amount?

10. ___5.0 mg___ Iodine-131 has a half-life of 8.0 days. How many grams of an original 160 mg sample will remain after 40 days?

11. ___0.24 mg___ Carbon-14 has a half-life of 5715 years. It is used to determine the age of ancient objects. If a sample today contains 0.060 mg of carbon-14, how much carbon-14 must have been present in the sample 11 430 years ago?

CHAPTER 21 REVIEW
Nuclear Chemistry

SECTION 3

SHORT ANSWER Answer the following questions in the space provided.

1. __d__ The radioisotope cobalt-60 is used for all of the following applications *except*
   - (a) killing food-spoiling bacteria.
   - (b) preserving food.
   - (c) treating heart disease.
   - (d) treating certain kinds of cancers.

2. __c__ All of the following contribute to background radiation exposure *except*
   - (a) radon in homes and buildings.
   - (b) cosmic rays passing through the atmosphere.
   - (c) consumption of irradiated foods.
   - (d) minerals in Earth's crust.

3. __b__ Which one of the graphs shown below best illustrates the decay of a sample of carbon-14? Assume each division on the time axis represents 5715 years.

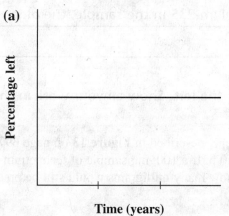
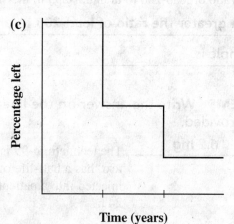
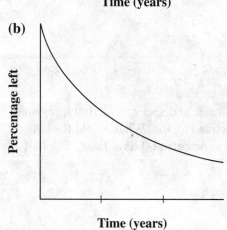
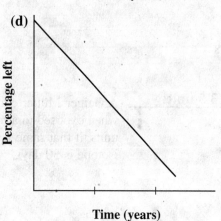

Name _____ Date _____ Class _____

SECTION 3 continued

4. Match the item on the left with its description on the right.

    __c__ Geiger-Müller counter (a) device that uses film to measure the approximate radiation exposure of people working with radiation

    __b__ scintillation counter (b) instrument that converts scintillating light to an electric signal for detecting radiation

    __a__ film badge (c) meter that detects radiation by counting electric pulses carried by gas ionized by radiation

    __d__ radioactive tracers (d) radioactive atoms that are incorporated into substances so that movement of the substances can be followed by detectors

5. Which type of radiation is easiest to shield? Why?

    Alpha radiation penetrates the least and is most easily absorbed.

6. One technique for dating ancient rocks involves uranium-235, which has a half-life of 710 million years. Rocks originally rich in uranium-235 will contain small amounts of its decay series, including the nonradioactive lead-206. Explain the relationship between a sample's relative age and the ratio of lead-206 to uranium-235 in the sample.

    The greater the ratio of lead-206 to uranium-235 in the sample, the older the rock sample is.

PROBLEMS Write the answer on the line to the left. Show all your work in the space provided.

7. __6.2 mg__ The technetium-99 isotope, described in **Figure 13** on page 697 of the text, has a half-life of 6.0 h. If a 100. mg sample of technetium-99 were injected into a patient, how many milligrams would still be present after 24 h?

8. __3.5 units__ A Geiger-Müller counter, used to detect radioactivity, registers 14 units when exposed to a radioactive isotope. What would the counter read, in units, if that same isotope is detected 60 days later? The half-life of the isotope is 30 days.

CHAPTER 21 REVIEW
Nuclear Chemistry

SECTION 4

SHORT ANSWER Answer the following questions in the space provided.

1. Match each of the following statements with the process(es) to which they apply, using one of the choices below:

   (1) fission only
   (2) fusion only
   (3) both fission and fusion
   (4) neither fission nor fusion

   __1__ a. A very large nucleus splits into smaller pieces.

   __3__ b. The total mass before a reaction is greater than the mass after a reaction.

   __1__ c. The rate of a reaction can be safely controlled for energy generation in suitable vessels.

   __2__ d. Two small nuclei form a single larger one.

   __3__ e. Less-stable nuclei are converted to more-stable nuclei.

2. Match the reaction type on the right to the statement(s) that applies to it on the left.

   __a, c__ It requires very high temperatures.   (a) uncontrolled fusion

   __d__ It is used in nuclear reactors to make electricity.   (b) uncontrolled fission

   __a__ It occurs in the sun and other stars.   (c) controlled fusion

   __b__ It is used in atomic bombs.   (d) controlled fission

3. Match the component of a nuclear power plant on the right to its use on the left.

   __c__ limits the number of free neutrons   (a) moderator

   __a__ is used to slow down neutrons   (b) fuel rod

   __f__ drives an electric generator   (c) control rod

   __b__ provides neutrons by its fission   (d) shielding

   __e__ removes heat from the system safely   (e) coolant

   __d__ prevents escape of radiation   (f) turbine

4. __b__ A chain reaction is any reaction in which

   (a) excess reactant is present.
   (b) the material that starts the reaction is also a product.
   (c) the rate is slow.
   (d) many steps are involved.

SECTION 4 continued

5. As a star ages, does the ratio of He atoms to H atoms in its composition become larger, smaller, or remain constant? Explain your answer.

   The ratio of He atoms to H atoms in a star's composition becomes larger. As a star ages, H atoms fuse into He atoms. A new star is almost 100% H, but the older the star becomes, the more of its H is converted into He.

6. The products of nuclear fission are variable; many possible nuclides can be created. In the feature "An Unexpected Finding," on page 702 of the text, it was noted that Meitner showed radioactive barium to be one product of fission. Following is an incomplete possible nuclear equation for the production of barium-141:

$$^{235}_{92}U + ^{1}_{0}n \rightarrow ^{141}_{56}Ba + \underline{\ ?\ } + 4\,^{1}_{0}n + \text{energy}$$

   $^{91}_{36}Kr$ a. Determine the missing fission product formed. Write the element's symbol, its atomic number, and its mass number.

   Yes   b. Is it likely that this isotope in part **a** is unstable? (Refer to **Figure 2** on page 685 of the text.)

7. Small nuclides can undergo fusion.

   $^{10}_{4}Be$ a. Complete the following nuclear equation by identifying the missing term. Write the element's symbol, its atomic number, and its mass number.

$$^{3}_{1}H + ^{7}_{3}Li \rightarrow \underline{\ ?\ } + \text{energy}$$

   b. When measured exactly, the total mass of the reactants does not add up to that of the products in the reaction represented in part **a**. Why is there a difference between the mass of the products and the mass of the reactants? Which has the greater mass, the reactants or the products?

   During the reaction, some of the mass is converted into energy. The mass of the reactants is greater than the mass of the products.

8. What are some current concerns regarding development of nuclear power plants?

   Current concerns include environmental requirements, safety of the operation, plant construction costs, and storage of spent fuel.

CHAPTER 21 REVIEW
Nuclear Chemistry

MIXED REVIEW

SHORT ANSWER Answer the following questions in the space provided.

1. The ancient alchemists dreamed of being able to turn lead into gold. By using lead-206 as the target atom of a powerful accelerator, modern chemists can attain that dream in principle. Write the nuclear equation for a one-step process that will convert $^{206}_{82}Pb$ into a nuclide of gold-79. You may use alpha particles, beta particles, positrons, or protons.

   $^{206}_{82}Pb + ^{0}_{-1}e \rightarrow ^{4}_{2}He + ^{202}_{79}Au$

2. A typical fission reaction releases 2×10^{10} kJ/mol of uranium-235, while a typical fusion reaction produces 6×10^{8} kJ/mol of hydrogen-1. Which process produces more energy from 235 g of starting material? Explain your answer.

   Fusion produces more energy per gram of starting material. For uranium-235, 235 g = 1.0 mol, and this releases 2×10^{10} J of energy in a fission reaction. For hydrogen-1, 235 g = 235 mol, and this releases $235(6 \times 10^{8}$ J) or 1.4×10^{11} J of energy in a fusion reaction. This is seven times more energy than from fission of uranium-235.

3. Write the nuclear equations for the following reactions:

   a. Carbon-12 combines with hydrogen-1 to form nitrogen-13.

   $^{12}_{6}C + ^{1}_{1}H \rightarrow ^{13}_{7}N$

   b. Curium-246 combines with carbon-12 to form nobelium-254 and four neutrons.

   $^{246}_{96}Cm + ^{12}_{6}C \rightarrow ^{254}_{102}No + 4^{1}_{0}n$

   c. Hydrogen-2 combines with hydrogen-3 to form helium-4 and a neutron.

   $^{2}_{1}H + ^{3}_{1}H \rightarrow ^{4}_{2}He + ^{1}_{0}n$

4. Write the complete nuclear equations for the following reactions:

   a. $^{91}_{42}Mo$ undergoes positron emission.

   $^{91}_{42}Mo \rightarrow ^{91}_{41}Nb + ^{0}_{+1}\beta$

   b. $^{6}_{2}He$ undergoes beta decay.

   $^{6}_{2}He \rightarrow ^{6}_{3}Li + ^{0}_{-1}\beta$

   c. $^{194}_{84}Po$ undergoes alpha decay.

   $^{194}_{84}Po \rightarrow ^{190}_{82}Pb + ^{4}_{2}He$

MODERN CHEMISTRY   NUCLEAR CHEMISTRY

MIXED REVIEW continued

d. $^{129}_{55}Cs$ undergoes electron capture.

$^{129}_{55}Cs + ^{0}_{-1}e \rightarrow ^{129}_{54}Xe$

PROBLEMS Write the answer on the line to the left. Show all your work in the space provided.

5. __2.2×10^{-12} J__ It was shown in **Section 1** of the text that a mass defect of 0.030 377 amu corresponds to a binding energy of 4.54×10^{-12} J. What binding energy would a mass defect of 0.015 amu yield?

6. __24 days__ Iodine-131 has a half-life of 8.0 days; it is used in medical treatments for thyroid conditions. Determine how many days must elapse for a 0.80 mg sample of iodine-131 in the thyroid to decay to 0.10 mg.

7. Following is an incomplete nuclear fission equation:

$$^{235}_{92}U + ^{1}_{0}n \rightarrow ^{90}_{38}Sr + ^{141}_{54}Xe + x\,^{1}_{0}n + \text{energy}$$

__5__ a. Determine the value of x in the above equation.

__$\frac{1}{8}$__ b. The strontium-90 produced in the above reaction has a half-life of 28 years. What fraction of strontium-90 still remains in the environment 84 years after it was produced in the reactor?

CHAPTER 22 REVIEW
Organic Chemistry

SECTION 1

SHORT ANSWER Answer the following questions in the space provided.

1. Name two types of carbon-containing molecules that are not organic.

   carbonates and oxides

2. __c__ Carbon atoms form bonds readily with atoms of

   (a) elements other than carbon.   (c) both carbon and other elements.
   (b) carbon only.   (d) only neutral elements.

3. Explain why the following two molecules are *not* geometric isomers of one another.

   Free rotation around the single bond between the carbon atoms will allow both of these configurations to occur with the same molecule.

4. a. In the space below, draw the structural formulas for two structural isomers with the same molecular formula.

   Answers will vary.

   b. In the space below, draw the structural formulas for two geometric isomers with the same molecular formula.

   Answers will vary.

SECTION 1 continued

5. Draw a structural formula that demonstrates the catenation of the methane molecule, CH_4.

   Answers will vary: any long hydrocarbon will do.

   $$\begin{array}{c} \text{H H H H H H} \\ | \; | \; | \; | \; | \; | \\ \text{H—C—C—C—C—C—C—H} \\ | \; | \; | \; | \; | \; | \\ \text{H H H H H H} \end{array}$$

6. Draw the structural formulas for two structural isomers of C_4H_{10}.

7. Draw the structural formula for the *cis*-isomer of $C_2H_2Cl_2$.

8. Draw the structural formula for the *trans*-isomer of $C_2H_2Cl_2$.

CHAPTER 22 REVIEW
Organic Chemistry

SECTION 2

SHORT ANSWER Answer the following questions in the space provided.

1. __a__ Hydrocarbons that contain only single covalent bonds between carbon atoms are called
   - (a) alkanes.
   - (b) alkenes.
   - (c) alkynes.
   - (d) unsaturated.

2. __c__ When the longest straight-chain in a hydrocarbon contains seven carbons, its prefix is
   - (a) pent-.
   - (b) hex-.
   - (c) hept-.
   - (d) oct-.

3. __b__ The alkyl group with the formula —CH_2—CH_3 is called
   - (a) methyl.
   - (b) ethyl.
   - (c) propyl.
   - (d) butyl.

4. What is a saturated hydrocarbon?

   a hydrocarbon in which each carbon atom forms four single covalent bonds with other atoms

5. Explain why the general formula for an alkane, C_nH_{2n+2}, correctly predicts hydrocarbons in a homologous series.

   Each nonterminal carbon atom within the hydrocarbon chain bonds with two hydrogen atoms. The two terminal carbon atoms on the chain bond with an additional hydrogen atom each to complete carbon's four covalent bonds.

6. Why is the general formula for cycloalkanes, C_nH_{2n}, different from the general formula for straight-chain hydrocarbons?

   There are no terminal carbon atoms requiring a third hydrogen in a cycloalkane.

Name _____ Date _____ Class _____

SECTION 2 continued

7. Write the IUPAC name for the following structural formulas:

   a. __3,4-diethylhexane__

   $$\begin{array}{c} \text{CH}_3-\text{CH}_2 \quad \text{CH}_2-\text{CH}_3 \\ | \quad\quad | \\ \text{CH}_3-\text{CH}_2-\text{CH}-\text{CH}-\text{CH}_2-\text{CH}_3 \end{array}$$

   b. __3-ethyl-3-methylpentane__

   $$\begin{array}{c} \text{CH}_3-\text{CH}_2 \\ | \\ \text{CH}_3-\text{CH}_2-\text{C}-\text{CH}_2-\text{CH}_3 \\ | \\ \text{CH}_3 \end{array}$$

   c. __2, 5, 6-trimethyloctane__

   $$\begin{array}{c} \text{CH}_3-\text{CH}-\text{CH}_2-\text{CH}_2-\text{CH}-\text{CH}-\text{CH}_2-\text{CH}_3 \\ | \quad\quad\quad\quad\quad\quad | \quad | \\ \text{CH}_3 \quad\quad\quad\quad\quad \text{CH}_3 \; \text{CH}_3 \end{array}$$

8. Draw the structural formula for each of the following compounds:

   a. 3,4-diethyl-2-methy-1-hexene

   $$\begin{array}{c} \text{CH}_3-\text{CH}_2 \quad \text{CH}_2-\text{CH}_3 \\ | \quad\quad | \\ \text{CH}_2=\text{C}-\text{CH}-\text{CH}-\text{CH}_2-\text{CH}_3 \\ | \\ \text{CH}_3 \end{array}$$

   b. 1-ethyl-2,3-dimethylbenzene

   (benzene ring with CH_2-CH_3, CH_3, CH_3 substituents)

   c. 5, 6-dimethyl-2-heptyne

   $$\text{CH}_3-\text{C}\equiv\text{C}-\text{CH}_2-\overset{\overset{\text{CH}_3}{|}}{\text{CH}}-\overset{\overset{\text{CH}_3}{|}}{\text{CH}}-\text{CH}_3$$

CHAPTER 22 REVIEW
Organic Chemistry

SECTION 3

SHORT ANSWER Answer the following questions in the space provided.

1. Match the structural formulas on the right to the family name on the left.

   __d__ aldehyde
   __f__ ketone
   __a__ carboxylic acid
   __b__ amine
   __c__ ester
   __e__ alkene

   (a) H—C—OH with =O below C
   (b) H—C—N—C—H with H's (amine structure)
   (c) H—C—O—C—H with =O below first C and H's
   (d) H—C—C=O with H's (aldehyde)
   (e) H₂C=CH₂ (alkene)
   (f) H—C—C—C—H with =O on middle C (ketone)

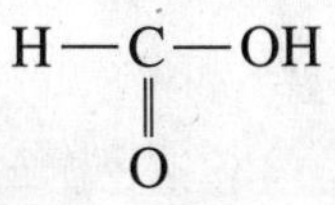

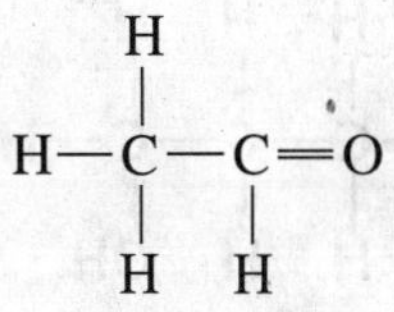

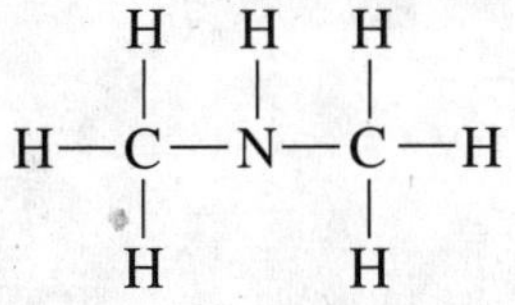

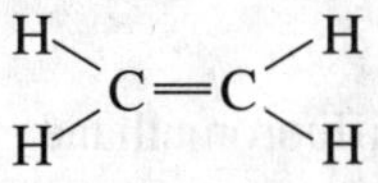

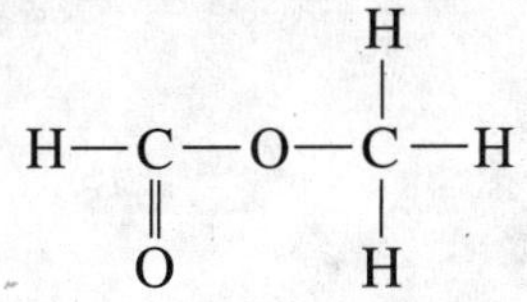

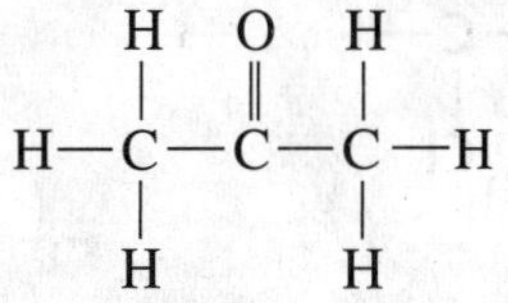

2. What is the functional group in glycerol? Explain how glycerol functions in skin care products.
   Glycerol is an alcohol containing three hydroxyl groups. This structure allows it to form multiple hydrogen bonds with water. Thus, glycerol functions as a moisturizer in skin care products.

3. List the halogen atoms found in alkyl halides in order of increasing atomic mass.
   fluorine, chlorine, bromine, iodine

4. State the difference between aldehydes and ketones.
   The difference is the location of the carbonyl group. In aldehydes, the carbonyl group is attached to a carbon atom at the end of a carbon-atom chain. In ketones, the carbonyl group is attached to carbon atoms within the chain.

SECTION 3 continued

5. __acetic acid__ Which is the weaker acid, acetic acid or sulfuric acid?

6. Explain why esters are considered derivatives of carboxylic acids.
   Esters are derivatives of carboxylic acids because of their structural similarity. The hydrogen atom of the acid's hydroxyl group is replaced by an alkyl group in esters.

7. Draw structural formulas for the following compounds:

   a. 1-butanol

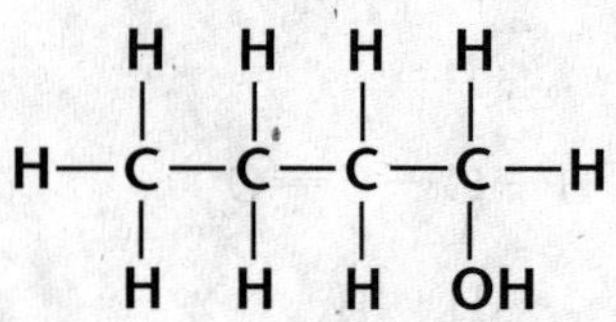

   $$\text{H-C(H)(H)-C(H)(H)-C(H)(H)-C(H)(OH)-H}$$

   b. dichlorodifluoromethane

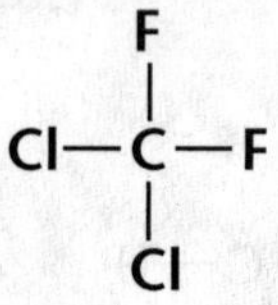

   $$\text{Cl-C(F)(F)-Cl}$$

184 ORGANIC CHEMISTRY

CHAPTER 22 REVIEW
Organic Chemistry

SECTION 4

SHORT ANSWER Answer the following questions in the space provided.

1. Match the reaction type on the left to its description on the right.

   __c__ substitution (a) An atom or molecule is added to an unsaturated molecule, increasing the saturation of the molecule.

   __a__ addition (b) A simple molecule is removed from adjacent atoms of a larger molecule.

   __d__ condensation (c) One or more atoms replace another atom or group of atoms in a molecule.

   __b__ elimination (d) Two molecules or parts of the same molecule combine.

2. Substitution reactions can require a catalyst to be feasible. The reaction represented by the following equation is heated to maximize the percent yield.

   $$C_2H_6(g) + Cl_2(g) + \text{energy} \overset{\Delta}{\rightleftharpoons} C_2H_5Cl(l) + HCl(g)$$

   __high__ a. Should a high or low temperature be maintained?

   __high__ b. Should a high or low pressure be used?

   __Yes__ c. Should the HCl gas be allowed to escape into another container?

3. Elemental bromine is a reddish-brown liquid. Hydrocarbon compounds that contain bromine are colorless. A qualitative test for carbon-carbon multiple bonds is the addition of a few drops of bromine solution to a hydrocarbon sample at room temperature and in the absence of sunlight. The bromine will either quickly lose its color or remain reddish brown.

   __addition__ a. If the sample is unsaturated, what type of reaction should occur when the bromine is added under the conditions mentioned above?

   __substitution__ b. If the sample is saturated, what type of reaction should occur when the bromine is added under the conditions mentioned above?

   __unsaturated__ c. The reddish brown color of a bromine solution added to a hydrocarbon sample at room temperature and in the absence of sunlight quickly disappears. Is the sample a saturated or unsaturated hydrocarbon?

SECTION 4 continued

4. Two molecules of glucose, $C_6H_{12}O_6$, undergo a condensation reaction to form one molecule of sucrose, $C_{12}H_{22}O_{11}$.

___1___ **a.** How many molecules of water are formed during this condensation reaction?

b. Write a balanced chemical equation for this condensation reaction.

$2C_6H_{12}O_6 \rightarrow C_{12}H_{22}O_{11} + H_2O$

5. Addition reactions with halogens tend to proceed rapidly and easily, with the two halogen atoms bonding to the carbon atoms connected by the multiple bond. Thus, only one isomeric product forms.

a. Write an equation showing the structural formulas for the reaction of Br_2 with 1-butene.

```
 H H H H                    H H H H
 | | | |                    | | | |
H-C=C-C-C-H + Br₂ →    H-C-C-C-C-H
     | |                    | | | |
     H H                    Br Br H H
```

b. Name the product.

1,2-dibromobutane

6. Identify each of the following substances as either a natural or a synthetic polymer.

___natural___ **a.** cellulose

___synthetic___ **b.** nylon

___natural___ **c.** proteins

7. The text gives several abbreviations commonly used in describing plastics or polymers. For each of the following abbreviations, give the full term and one common household usage.

a. HDPE

high-density polyethylene; rigid plastic bottles

b. LDPE

low-density polyethylene; plastic shopping bags

c. cPE

cross-linked polyethylene; plastic crates

8. Explain why an alkane cannot be used as the monomer of an addition polymer.

There must be a double or triple bond onto which an adjoining CH_2 group can add.

An alkane has all single bonds.

CHAPTER 22 REVIEW
Organic Chemistry

MIXED REVIEW

SHORT ANSWER Answer the following questions in the space provided.

1. __a__ A saturated organic compound
   (a) contains all single bonds.
   (b) contains at least one double or triple bond.
   (c) contains only carbon and hydrogen atoms.
   (d) is quite soluble in water.

2. Arrange the following in order of increasing boiling point:

   __1__ a. ethane

   __2__ b. pentane

   __3__ c. heptadecane

3. Recall that isomers in organic chemistry have identical molecular formulas but different structures and IUPAC names.

   __True__ a. Two isomers must have the same molar mass. True or False?

   __False__ b. Two isomers must have the same boiling point. True or False?

4. Explain why hydrocarbons with only single bonds cannot form geometric isomers.
   The free rotation around the single bonds between carbon atoms prevents molecules with the same sequence of atoms from having different orientations in space.

5. Write the IUPAC name for the following structural formulas:

   a. __4-ethyl-3,5-dimethylheptane__

   $CH_3-CH_2 \quad CH_2-CH_3$
   $\quad\quad\quad | \quad\quad\quad |$
   $CH_3-CH-CH-CH-CH_2-CH_3$
   $\quad\quad\quad\quad\quad\quad |$
   $\quad\quad\quad\quad\quad\quad CH_3$

   b. __3-methyl-1-butene__

   $\quad\quad CH_3$
   $\quad\quad |$
   $CH_3-CH-CH=CH_2$

MODERN CHEMISTRY — ORGANIC CHEMISTRY

MIXED REVIEW continued

__1,2,4-trichlorobutane__   c.

$$\text{H}-\underset{\underset{\text{H}}{|}}{\overset{\overset{\text{Cl}}{|}}{\text{C}}}-\underset{\underset{\text{H}}{|}}{\overset{\overset{\text{Cl}}{|}}{\text{C}}}-\underset{\underset{\text{H}}{|}}{\overset{\overset{\text{H}}{|}}{\text{C}}}-\underset{\underset{\text{H}}{|}}{\overset{\overset{\text{Cl}}{|}}{\text{C}}}-\text{H}$$

6. Draw the structural formula for each of the following compounds:

   a. 1,2,4-trimethylcyclohexane

   b. 3-methyl-1-pentyne CH≡C—CH—CH₂—CH₃
                                      |
                                      CH₃

7. Each of the following names implies a structure but is not a correct IUPAC name. For each example, draw the implied structural formula and write the correct IUPAC name.

   a. 3-bromopropane

   1-bromopropane

   b. 3,4-dichloro-4-pentene

   2,3-dichloro-1-pentene

8. Match the general formula on the right to the corresponding family name on the left.

   __e__ carboxylic acid   (a) R—X   (f) R—N—R″
                                                                  |
   __h__ ester   R′
                                        O
                                        ‖
   __d__ alcohol   (b) R—C—H   (g)   O
                                                                ‖
   __c__ ether   R—C—R′
                               (c) R—O—R′
   __a__ alkyl halide
                               (d) R—OH
   __f__ amine
                                                               O
                                        O   ‖
   __b__ aldehyde   (e) R—C—OH   (h) R—C—O—R′

   __g__ ketone

Name _____ Date _____ Class _____

CHAPTER 23 REVIEW
Biological Chemistry

SECTION 1

SHORT ANSWER Answer the following questions in the space provided.

1. __c__ Lactose and sucrose are both examples of
   - (a) lipids.
   - (b) monosaccharides.
   - (c) disaccharides.
   - (d) proteins.

2. __c__ Carbohydrates made up of long chains of glucose units are called
   - (a) monosaccharides.
   - (b) disaccharides.
   - (c) polysaccharides.
   - (d) simple sugars.

3. __c__ The disaccharide that is commonly known as table sugar is
   - (a) lactose.
   - (b) fructose.
   - (c) sucrose.
   - (d) maltose.

4. __a__ The polysaccharide that plants use for storing energy is
   - (a) starch.
   - (b) glycerol.
   - (c) cellulose.
   - (d) glycogen.

5. __b__ Many animals store carbohydrates in the form of
   - (a) starch.
   - (b) glycogen.
   - (c) cellulose.
   - (d) glycerol.

6. __d__ Which class of biomolecules includes fats, oils, waxes, steroids, and cholesterol?
   - (a) starches
   - (b) monosaccharides
   - (c) disaccharides
   - (d) lipids

Name _____ Date _____ Class _____

SECTION 1 continued

7. Relate the structure of carbohydrates to their role in biological systems.

   Carbohydrates are often used by living organisms to provide energy. Some carbohydrates provide rigid structure to plants and animals. Glucose, a monosaccharide, is the chemical that the bloodstream uses to carry energy throughout the body. Sugars can join together to form polysaccharides which can store energy. These processes are the main ways that living organisms use and store energy.

8. What is a condensation reaction, what is a hydrolysis reaction, and how do they differ?

   A condensation reaction is a reaction in which two molecules or parts of the same molecule combine, releasing water. A hydrolysis reaction is a chemical reaction in which water reacts with another substance to form two or more new substances. Condensation reactions generally build larger molecules from smaller ones, whereas hydrolysis reactions break down large molecules into smaller molecules.

9. Why can cows digest cellulose, while humans cannot?

   The glucose molecules in cellulose chains form hydrogen bonds that link the hydroxyl group of the glucose molecules to form insoluble, tough, fibrous sheets which humans cannot digest. Cows have extra stomachs that can hold the cellulose for a long time while microorganisms break down the cellulose into glucose.

10. Describe how phospholipids are arranged in the cell membrane.

    Phospholipids are arranged in a bilayer at the surface of the cell. The hydrophilic heads are on the outside surfaces of the bilayer. These charged heads are in contact with water-containing solutions both inside the cell and surrounding the cell. The hydrophobic tails point into the membrane, away from the water-containing solutions, so the membrane forms a boundary between the inside of the cell and its external environment.

CHAPTER 23 REVIEW
Biological Chemistry

SECTION 2

SHORT ANSWER Answer the following questions in the space provided.

1. __d__ Proteins are polypeptides made of many
    - (a) lipids.
    - (b) carbohydrates.
    - (c) starches.
    - (d) amino acids.

2. __d__ The side chains of amino acids may contain
    - (a) acidic and basic groups.
    - (b) polar groups.
    - (c) nonpolar groups.
    - (d) All of the above

3. __a__ The amino acid sequence of a polypeptide chain is its
    - (a) primary structure.
    - (b) secondary structure.
    - (c) tertiary structure.
    - (d) quaternary structure.

4. __c__ The secondary structure of a protein that is shaped like a coil, with hydrogen bonds that form along a single segment of peptide, is
    - (a) a looped structure.
    - (b) the active site.
    - (c) an alpha helix.
    - (d) a beta pleated sheet.

5. According to the text, which amino acid(s) contains a side chain

    __cysteine__ a. in which molecules form covalent disulfide bridges with each other?

    __valine__ b. that is hydrophobic?

    __asparagine or glutamic acid__ c. that forms hydrogen bonds?

    __histidine__ d. that is basic?

6. Identify each protein structure level described below.

   __tertiary structure__ a. may involve hydrogen bonds, salt bridges, and disulfide bonds that determine a protein's three-dimensional structure

   __quaternary structure__ b. is determined by the interaction of several polypeptides coming together

   __primary structure__ c. is the amino acid sequence of a protein

7. How many different tripeptides can be formed from one molecule of glycine and two molecules of valine? Draw the isomers using their three-letter codes.

   Three: Gly-Val-Val, Val-Gly-Val, Val-Val-Gly

8. What are the functions of fibrous proteins?

   Fibrous proteins are insoluble in water and serve as support filaments, cables, or sheets to give strength or protection to structures of living things. Examples are keratin, collagen, fibrin, elastins, and myosins.

9. How do enzymes work?

   Enzymes are catalysts that speed up metabolic reactions without being permanently changed or destroyed. Enzymes lower the activation energy required to carry out the reaction. They can be thought of as a lock and key: only an enzyme of a specific shape can fit the reactants in the reaction that it is speeding up.

CHAPTER 23 REVIEW
Biological Chemistry

SECTION 3

SHORT ANSWER Answer the following questions in the space provided.

1. __b__ The primary energy exchange in the body is the cycle between

    (a) amino acids and proteins.
    (b) ATP and ADP.
    (c) lipids and carbohydrates.
    (d) DNA and RNA.

2. __c__ In which equation is the hydrolysis reaction of ATP represented?

    (a) $ADP^{3-}(aq) + H_2O(l) \rightarrow ATP^{4-}(aq) + H_2PO_4^{-}(aq)$
    (b) $ATP^{4-}(aq) + H_2PO_4^{-}(aq) \rightarrow ADP^{3-}(aq) + H_2O(l)$
    (c) $ATP^{4-}(aq) + H_2O(l) \rightarrow ADP^{3-}(aq) + H_2PO_4^{-}(aq)$
    (d) None of the above

3. __a__ In the citric acid cycle,

    (a) CO_2 and ATP are formed.
    (b) food is digested.
    (c) glucose is formed.
    (d) DNA is replicated.

4. __b__ Animals can produce ATP molecules in

    (a) photosynthesis.
    (b) the Krebs cycle.
    (c) peptide synthesis.
    (d) DNA replication.

5. __c__ In glucogenesis, glucose is synthesized from

    (a) sucrose and fructose.
    (b) water and amino acids.
    (c) lactate, pyruvate, glycerol, and amino acids.
    (d) DNA and RNA.

Name _____ Date _____ Class _____

6. Identify each function as that of *autotrophs* or of *heterotrophs*.

   a. __autotrophs__ synthesize carbon-containing biomolecules from H_2O and CO_2

   b. __autotrophs__ absorb solar energy, which is converted into ATP

   c. __heterotrophs__ obtain energy by consuming plants or animals

7. Identify each as either a *catabolic process* or an *anabolic process*.

   a. Synthesis of protein molecules is a(n) __anabolic process__.

   b. A(n) __catabolic process__ releases energy.

   c. Digestion is a(n) __catabolic process__.

   d. A(n) __anabolic process__ requires energy.

8. How do plants use photosynthesis to gather energy?
   Most plants use chlorophyll to capture energy from sunlight. This energy is immediately converted into two energy-containing compounds, ATP and NADPH, which are used to produce carbohydrates. The carbohydrates can be used for energy or stored for later use.

9. Explain how animals indirectly gather energy from the sun.
   Animals eat plants, which make carbohydrates that animals, too, can use for energy. Once an animal eats a plant, it breaks down some of the plant's large carbohydrates into simpler carbohydrates, such as glucose. Glucose can be carried throughout the animal's body in the bloodstream to provide energy for the animal's activities.

CHAPTER 23 REVIEW
Biological Chemistry

SECTION 4

SHORT ANSWER Answer the following questions in the space provided.

1. Complete the following statements with *DNA* or *RNA*.

   a. _____DNA_____ is most often found in the form of a double helix.

   b. _____RNA_____ contains ribose as its sugar unit.

   c. _____RNA_____ is most often single stranded.

   d. _____RNA_____ is directly responsible for the synthesis of proteins.

2. Complete the following statements with the name of the correct base. More than one answer may be used.

   a. ____T, C, or U____ contains a six-membered ring called a pyrimidine.

   b. _____U_____ is the complementary base of A in RNA.

   c. ____A or G____ contains a five-membered ring called a purine.

   d. _____C_____ is the complementary base of G in DNA.

3. A segment of DNA has the base sequence TAC TTT TCG AAG AGT ATT.

   a. What is the base sequence in a complementary strand of RNA?

   AUG AAA AGC UUC UCA UAA

   b. What is the base sequence in a complementary strand of DNA?

   ATG AAA AGC TTC TCA TAA

4. A segment of DNA has the base sequence TAC CTT ACA GAT TGT ACT.

   a. What is the base sequence in a complementary strand of RNA?

   AUG GAA UGU CUA ACA UGA

   b. What is the base sequence in a complementary strand of DNA?

   ATG GAA TGT CTA ACA TGA

SECTION 4 continued

5. Explain why, in DNA, pairing exists only between A and T and between C and G?

   The pairing exists because the large purine (A or G) and the small pyrimidine (T or C) provide the correct width for the space between the two sides of the DNA ladder and they can hydrogen bond correctly. There is not enough room for two purines to fit, and two pyrimidines would be too far apart from each other to hydrogen bond effectively.

6. What is cloning, and how has it been accomplished in mammals?

   Cloning is producing an offspring that is genetically identical to a parent. It occurs naturally when identical twins result from a chance splitting of early embryonic cells. Artificial cloning, using stem cells from animals or meristem cells from plants, can be used to produce identical replicas of an organism.

CHAPTER 23 REVIEW
Biological Chemistry

MIXED REVIEW

SHORT ANSWER Answer the following questions in the space provided.

1. Use *carbohydrate(s)*, *lipid(s)*, *protein(s)*, or *DNA* to complete the following statements.

    a. _____DNA_____ is the largest molecule found in cells.

    b. _____Lipids_____ are the major component in a cell membrane.

    c. _____Carbohydrates_____ provides most of the energy that is available in plant-derived food.

    d. _____Protein_____ gets its name from the Greek word meaning "of first importance."

2. Describe four different kinds of interactions between side chains on a polypeptide molecule that help to make the shape that a protein takes.

    1. Disulfide bridges (covalent bonds) between side chains can form a looped protein or bond two separate polypeptides. 2. Ionic bonds can link different points on a protein. 3. A hydrophobic environment attracts other nonpolar molecules or nonpolar segments of the same protein. 4. Hydrogen bonds can form to oxygen atoms, especially at carboxyl groups.

3. How does DNA replicate itself?

    DNA replicates by separating its double helix into two strands. Each original strand acts as a template for making a new strand. Each base forms a hydrogen bond to its complementary base (A to T, T to A, C to G, G to C). Eventually, there are two identical double strands.

MIXED REVIEW continued

4. Draw the reaction of ATP hydrolysis to ADP, indicating the free energy.

   $ATP^{4-}(aq) + H_2O(l) \rightarrow ADP^{3-}(aq) + H_2PO_4^{-}(aq)$ $\Delta G = -31$ kJ

5. For the following peptide molecule, identify the peptide linkages, the amino groups, and the carboxyl groups.

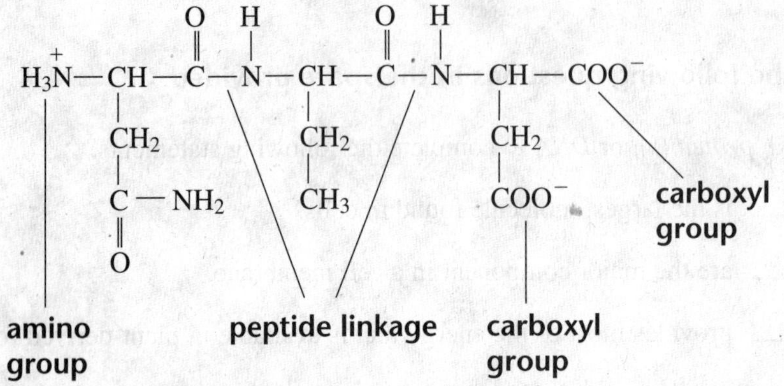

6. A segment of RNA has the base sequence UAG CCU AAG CGA UAC GGC ACG.

   a. What is the base sequence in a complementary strand of RNA?

   AUC GGA UUC GCU AUG CCG UGC

   b. What is the base sequence in the complementary strand of DNA?

   ATC GGA TTC GCT ATG CCG TGC.

7. Draw the condensation reaction of two molecules of glucose.glucose.

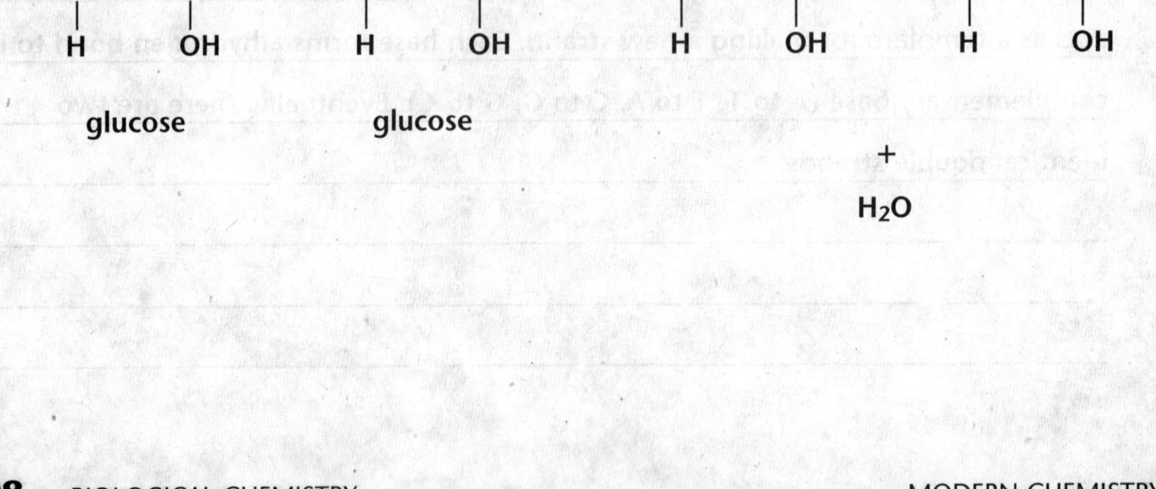